Communications in Computer and Information Science 3127

Rationale

The CCIS series is devoted to the publication of proceedings of computer science conferences. Its aim is to efficiently disseminate original research results in informatics in printed and electronic form. While the focus is on publication of peer-reviewed full papers presenting mature work, inclusion of reviewed short papers reporting on work in progress is welcome, too. Besides globally relevant meetings with internationally representative program committees guaranteeing a strict peer-reviewing and paper selection process, conferences run by societies or of high regional or national relevance are also considered for publication.

Topics

The topical scope of CCIS spans the entire spectrum of informatics ranging from foundational topics in the theory of computing to information and communications science and technology and a broad variety of interdisciplinary application fields.

Information for Volume Editors and Authors

Publication in CCIS is free of charge. No royalties are paid, however, we offer registered conference participants temporary free access to the online version of the conference proceedings on SpringerLink (http://link.springer.com) by means of an http referrer from the conference website and/or a number of complimentary printed copies, as specified in the official acceptance email of the event.

CCIS proceedings can be published in time for distribution at conferences or as post-proceedings, and delivered in the form of printed books and/or electronically as USBs and/or e-content licenses for accessing proceedings at SpringerLink. Furthermore, CCIS proceedings are included in the CCIS electronic book series hosted in the SpringerLink digital library at http://link.springer.com/bookseries/7899. Conferences publishing in CCIS are allowed to use our online conference service (Meteor) for managing the whole proceedings lifecycle (from submission and reviewing to preparing for publication) free of charge.

Publication process

The language of publication is exclusively English. Authors publishing in CCIS have to sign the Springer CCIS copyright transfer form, however, they are free to use their material published in CCIS for substantially changed, more elaborate subsequent publications elsewhere. For the preparation of the camera-ready papers/files, authors have to strictly adhere to the Springer CCIS Authors' Instructions and are strongly encouraged to use the CCIS LaTeX style files or templates.

Abstracting/Indexing

CCIS is abstracted/indexed in DBLP, Google Scholar, EI-Compendex, Mathematical Reviews, SCImago, Scopus. CCIS volumes are also submitted for the inclusion in ISI Proceedings.

How to start

To start the evaluation of your proposal for inclusion in the CCIS series, please send an e-mail to ccis@springer.com

Shehroz S. Khan · Luca Romeo · Nour Moustafa
Editors

AI for Aging, Rehabilitation and Intelligent Assisted Living

The 9th Workshop ARIAL@IJCAI, IJCAI 2026
Bremen, Germany, August 17, 2026
Proceedings

Springer

Editors
Shehroz S. Khan
American University of the Middle East
Egaila, Kuwait

Luca Romeo
University of Macerata
Macerata, Italy

Nour Moustafa
American University of the Middle East
Egaila, Kuwait

ISSN 1865-0929 ISSN 1865-0937 (electronic)
Communications in Computer and Information Science
ISBN 978-981-92-5189-6 ISBN 978-981-92-5190-2 (eBook)
https://doi.org/10.1007/978-981-92-5190-2

This Springer imprint is published by the registered company Springer Nature Singapore Pte Ltd.
The registered company address is: 152 Beach Road, #21-01/04 Gateway East, Singapore 189721, Singapore

Preface

This volume presents the proceedings of the 9th Workshop on AI for Aging, Rehabilitation, and Intelligent Assisted Living (ARIAL 2026), held in conjunction with the International Joint Conference on Artificial Intelligence and the European Conference on Artificial Intelligence (IJCAI-ECAI 2026) in Bremen, Germany, on August 17th, 2026.

The ARIAL workshop has established itself as a forum for researchers and practitioners working at the intersection of artificial intelligence, aging, rehabilitation, and intelligent assisted living. The workshop aims to foster interdisciplinary discussions on novel machine learning and artificial intelligence methodologies for monitoring, assessment, rehabilitation, and clinical decision support in aging-related applications. Topics of interest include multimodal data analysis, wearable and ambient sensing, digital health technologies, trustworthy and explainable AI, privacy-preserving machine learning, foundation models, large language models, and intelligent assistive systems that promote healthy and independent aging.

This year's edition attracted the largest number of submissions in the history of the workshop. A total of 19 papers were submitted and underwent a rigorous double-blind peer-review process. Each submission was reviewed by at least two independent reviewers, ensuring a fair and thorough evaluation based on scientific quality, originality, technical soundness, and relevance to the workshop themes. Following the review process, 12 full papers were accepted for publication in this volume, corresponding to an acceptance rate of 63.2%.

We sincerely thank all authors for submitting their work and the members of the Program Committee for their valuable time, constructive feedback, and commitment to maintaining the scientific quality of the workshop. We are also grateful to the IJCAI-ECAI 2026 Workshop Chairs and the Springer CCIS editorial team for their continued support throughout the organization of the workshop and the preparation of these proceedings.

We hope that these proceedings will provide a valuable resource for researchers and practitioners working on artificial intelligence methods for aging, rehabilitation, and intelligent assisted living, and will stimulate further advances in this rapidly evolving interdisciplinary field.

August 2026

Shehroz S. Khan
Luca Romeo
Nour Moustafa

Organization

Program Committee Chairs

Shehroz S. Khan	American University of the Middle East, Kuwait
Luca Romeo	University of Macerata, Italy
Nour Moustafa	American University of the Middle East, Kuwait

Program Committee

Ali Abedi	University Health Network, Canada
Michele Bernardini	Università degli Studi eCampus, Italy
Veselka Boeva	Blekinge Institute of Technology, Sweden
Reza Basiri	University of Toronto, Canada
Mariachiara Di Cosmo	Marche Polytechnic University, Italy
Pratik Kumar Mishra	University of Toronto, Canada
Ali Nawaz	United Arab Emirates University, United Arab Emirates
Riccardo Rosati	University of Macerata, Italy
Alessio Giorgetti	University of Pisa, Italy
Maitrik Patel	Apple Inc., USA
Maria Chiara Fiorentino	Università degli Studi "Gabriele d'Annunzio" Chieti, Italy
Shahrooz Abghari	Blekinge Institute of Technology, Sweden

Contents

AI for Rehabilitation and Clinical Decision Support

MAISON-LLF Data Challenge

The MAISON Generalization Challenge: Participant-Aware Evaluation Reveals Limited Cross-Subject Generalization in Recovery Monitoring

Alessio Giorgetti[1], Riccardo Rosati[2], and Luca Romeo[3(✉)]

[1] Department of Management, Università Politecnica delle Marche, Ancona, Italy
a.giorgetti@staff.univpm.it

[2] Department of Political Sciences, Communication and International Relations, Università degli Studi di Macerata, Macerata, Italy
riccardo.rosati@unimc.it

[3] Department of Economics and Law, Università degli Studi di Macerata, Macerata, Italy
luca.romeo@unimc.it

Abstract. Lower-limb fractures in older adults are often associated with prolonged functional decline, reduced mobility, and increased social isolation during post-surgical recovery. Recent advances in multimodal passive sensing provide new opportunities for continuously monitoring behavioral and recovery patterns in home environments. However, robust predictive modeling remains challenging because of substantial variability across individuals and the limited size of longitudinal clinical datasets. In this work, we introduce the MAISON-LLF Data Challenge, originating from the MAISON multimodal dataset, to investigate the feasibility of predicting clinically relevant social and functional recovery outcomes from passive sensing data. The analyzed dataset includes 18 participants, 1,008 day-level observations, and 72 biweekly clinical assessments collected from smartwatch, smartphone, motion, sleep, and demographic data sources. To evaluate generalization under realistic deployment conditions, we adopted a strict nested participant-aware validation framework combined with participant-aware hyperparameter optimization. Multiple machine learning models were evaluated to predict recovery outcomes from multimodal sensing features. The results reveal that predicting functional recovery and social isolation trajectories across unseen participants remains highly challenging. Most approaches exhibited unstable predictive performance and limited explanatory power, highlighting the difficulty of learning transferable behavioral representations under substantial participant heterogeneity. Simpler models demonstrated more stable behavior than more complex approaches, suggesting a reduced tendency to overfit participant-specific patterns under limited-data conditions. Overall, this study emphasizes the importance of rigorous participant-aware evaluation in multimodal digital health research and identifies cross-subject generalization as a major open challenge for

S. S. Khan et al. (Eds.): IJCAI 2026, CCIS 3127, pp. 3–13, 2027.
https://doi.org/10.1007/978-981-92-5190-2_1

passive monitoring systems supporting recovery after lower-limb fractures.

Keywords: Multimodal sensing · Passive monitoring · Cross-subject generalization · Digital health · Lower-limb fractures · Machine learning

1 Introduction

Lower-limb fractures (LLF) represent a severe clinical challenge within aging populations, often inducing a prolonged loss of mobility that heavily compromises autonomy, social participation, and overall quality of life [4]. Following surgical intervention, older adults typically undergo the majority of their rehabilitation at home. In these community settings, the absence of continuous clinical oversight creates a critical vulnerability window where secondary conditions such as social isolation and accelerated functional decline frequently manifest undetected [8]. Traditional clinical workflows rely heavily on sporadic, patient-initiated self-reports or infrequent in-clinic checkups, which are prone to subjective bias, social stigma, and a lack of real-time granularity. Consequently, there is an urgent need for passive, non-invasive digital health platforms capable of continuously monitoring patient recovery trajectories to alert clinicians before irreversible physical or psychological deterioration occurs [10].

Recent advancements in multimodal sensing and internet-of-things (IoT) ecosystems have opened promising avenues for automatic, remote behavioral assessment. By leveraging passive data stream combinations—such as smartwatch inertial metrics, smartphone-based geographical position tracking, and ambient motion telemetry—machine learning paradigms can be engineered to passively model a patient's daily life attributes [2]. While early frameworks demonstrated the feasibility of detecting behavioral deviations through basic correlation studies, deploying robust predictive models in real-world clinical environments remains constrained by data limitations. Most existing datasets suffer from small sample sizes, brief observational windows, or a single-modality focus [11]. The initial release of the Multimodal AI-Based Sensor Platform for Older Individuals dataset (MAISON-LLF v1 [1]) took an important step forward by releasing 560 days of continuous 24-hour sensor data paired with biweekly clinical gold-standard questionnaires from 10 participants.

Recent advances in foundation models for tabular data [6] and deep sequential learning architectures have expanded the range of approaches available for analyzing multimodal health data. Such methods have demonstrated considerable potential for biomedical anomaly detection [3], activity recognition [9], rehabilitation monitoring [7], and longitudinal modeling of behavioral and recovery trajectories in older adults. However, before investigating increasingly sophisticated predictive architectures, it is important to understand whether the underlying sensing features exhibit sufficient cross-subject transferability. Consequently, the focus of this study is on participant-aware evaluation and the characterization of inter-subject variability rather than on maximizing predictive performance.

This requires both a sufficiently rich multimodal dataset and an evaluation protocol capable of realistically assessing model performance on previously unseen individuals.

To address these challenges, this work utilizes an extended iteration of the MAISON-LLF dataset[1], expanding the cohort to 18 participants and encompassing 1,008 day-level observations. Building upon this larger dataset, we evaluate a diverse set of machine learning approaches—including tree-based, ensemble, and kernel-based models—to predict the Social Isolation Scale (SIS) and the Oxford Knee Score (OKS) from daily passive sensing data. Crucially, all models are assessed through a strict participant-aware validation framework designed to quantify the impact of inter-subject variability and the ability of learned representations to generalize across unseen participants.

The main contributions of this work are twofold:

- we establish a rigorous participant-aware evaluation protocol for assessing predictive models on the MAISON-LLF dataset, an extension of the MAISON multimodal sensing dataset for older adults recovering from LLF, under realistic deployment conditions. Specifically, all models are evaluated using a strict nested Leave-One-Participant-Out (LOPO) framework that prevents participant-level data leakage during both hyperparameter optimization and final evaluation;
- we demonstrate that participant heterogeneity constitutes a major obstacle to cross-subject generalization in the MAISON-LLF dataset. Despite the use of multiple machine learning approaches and participant-aware model selection, the obtained results reveal limited transferability of behavioral representations across unseen participants, highlighting important challenges for passive monitoring of social and functional recovery trajectories in older adults.

Rather than aiming to maximize predictive accuracy, this study provides a realistic assessment of the challenges associated with predicting clinical recovery outcomes from passive multimodal sensing data.

2 Materials and Methods

2.1 Dataset and Clinical Outcomes

The MAISON-LLF dataset combines smartphone, smartwatch, motion, step, and sleep-tracking data from older adults recovering from LLF with clinical questionnaires on social isolation and functional decline collected every two weeks, in particular the SIS and OKS. The SIS is a six-item instrument where higher values indicate more social interaction and lower isolation, and the OKS scores physical-function and pain-related level where higher values indicate better status. The dataset is composed of two CSV files: a feature table and a demographic table. The feature table contains 1,008 daily observations from 18 participants, with exactly 56 monitored days per participant. Each participant has four clinical

[1] https://doi.org/10.5281/zenodo.17943110.

assessment dates, yielding 72 participant-assessment records. The demographic table contains sex, age, fracture type, relationship status, education, work status, and ethnicity. Participants had a mean age of 76.5 years (standard deviation 8.6; range 60 to 94), with 14 female and 4 male participants. Fracture categories included hip replacement, hip fracture, pelvis fracture, and femur fracture. The target outcomes in the regression analysis were SIS and OKS. SIS ranged from 15 to 30 in the day-level table, with a mean of 23.7 and standard deviation of 3.8. OKS ranged from 15 to 48, with a mean of 30.2 and standard deviation of 9.7. OHS was temporarily excluded from the analysis. Daily sensor features covered acceleration, heart rate, indoor motion, outdoor position, sleep, and step modalities.

2.2 Preprocessing

The preprocessing pipeline integrated the daily multimodal feature dataset with participant-level demographic information before model training. Categorical demographic variables were then transformed into numerical representations: sex was encoded as a binary variable, with male coded as 1 and female as 0; work status was encoded by distinguishing retired participants from those employed part-time; and education level was converted into an ordinal variable ranging from secondary education to doctorate degree. Nominal variables with no intrinsic ordering, namely fracture type and ethnicity, were transformed using one-hot encoding. Only numerical variables were retained as input features for the machine learning models. The final predictor matrix contained 57 variables: 16 acceleration-derived features, 6 heart-rate features, 5 motion features, 3 position features, 11 sleep features, 5 step features, and 11 demographic features.

2.3 Machine Learning Models and Validation

The final machine learning pipeline was formulated exclusively as a supervised regression problem, with the clinical scores treated as continuous outcomes. Specifically, the models were trained to predict the SIS and the OKS. The regression models evaluated in the comparative framework included XGBoost Regressor, LightGBM Regressor, Decision Tree Regressor (DT), and Support Vector Regressor (SVR). These models were selected to compare different families of tabular machine learning approaches, including gradient-boosting ensemble methods, tree-based regression, and kernel-based regression. For XGBoost and LightGBM, the explored hyperparameters included the number of estimators, maximum tree depth, and learning rate; for the DT model, the maximum depth was optimized; and for SVR, where the radial basis function was used as kernel, the regularization parameter (C), epsilon-insensitive loss margin, and gamma setting were tuned.

Model selection was performed using a nested participant-aware validation strategy. In the outer loop, a LOPO validation scheme was adopted, where all samples from one participant were held out as the independent test set. Within each outer training set, hyperparameter optimization was carried out using an inner LOPO procedure, in which each training participant was iteratively used as the validation subject. The optimal hyperparameter configuration was selected maximizing the mean validation coefficient of determination calculated across the inner folds. The selected model was then refitted on the entire outer training set and evaluated on the held-out participant. Regression performance was assessed using multiple complementary metrics, including mean absolute error (MAE), mean squared error (MSE), root mean squared error (RMSE), coefficient of determination (R^2), mean absolute percentage error (MAPE), median absolute error (MEDAE), and explained variance score (EVS). This nested LOPO design ensured that both hyperparameter tuning and final model evaluation were performed without participant-level data leakage, providing a rigorous estimate of model generalization performance. It is important to note that the coefficient of determination (R^2) was computed using the standard implementation provided by the `scikit-learn` library. Under this formulation, negative R^2 values naturally occur whenever the prediction error becomes larger than the variability explained by a naive mean-based predictor.

2.4 Hyperparameter Optimization and Code Availability

The hyperparameter search spaces explored for each regression model are summarized in Table 1. Hyperparameter optimization was conducted independently within each inner LOPO fold using a nested LOPO procedure to ensure complete participant separation between model selection and final evaluation, thereby preventing participant-level data leakage throughout the entire validation pipeline.

The source code and scripts used to perform the analyses are publicly available at: https://github.com/a-giorgetti-univpm/MAISON-ML/tree/main.

Table 1. Hyperparameter configurations considered during the INNER-LOPO hyperparameter optimization procedure for the evaluated regression approaches.

Model	Parameter	Values
XGBoost	Learning rate η	$\{0.01, 0.001\}$
	Maximum depth of the trees D_{max}	$\{5, 10, 20, 50\}$
	Number of estimators T	$\{20, 50\}$
LightGBM	Learning rate η	$\{0.01, 0.001\}$
	Maximum depth of the trees D_{max}	$\{5, 10, 20, 50\}$
	Number of estimators T	$\{20, 50\}$
DT	Maximum depth of the trees D_{max}	$\{5, 10, 20, 50\}$
SVR	Kernel	{`rbf`}
	Regularization parameter C	$\{1, 10\}$
	Epsilon ϵ	$\{0.01, 0.2\}$
	Kernel coefficient γ	{`scale`}

3 Results

Table 2 summarizes the average values and standard deviations of the regression evaluation metrics obtained across the nested LOPO folds for all evaluated models and both target variables. Overall, the results reveal limited predictive generalization capability across participants, with most approaches exhibiting strongly negative R^2 values and large variability across folds. These findings indicate that predicting clinical recovery outcomes from passive multimodal sensing data under strict participant-aware validation remains highly challenging, suggesting limited transferability of behavioral representations across unseen individuals. Notably, these observations are consistent with previous experiments conducted on earlier versions of the MAISON-LLF dataset [1], where substantial inter-subject variability and limited cross-participant generalization were similarly identified as major challenges for predictive modeling in passive recovery monitoring scenarios.

Importantly, the instability observed for several boosting-based and kernel-based approaches persisted despite the nested participant-aware hyperparameter optimization procedure summarized in Table 1. This suggests that the limited predictive performance primarily arises from the intrinsic difficulty of modeling highly heterogeneous recovery trajectories rather than from suboptimal hyperparameter selection.

Table 2. Average values and standard deviations (Mean ± SD) of regression metrics for SIS and OKS targets.

Target	Model	MAE	MSE	RMSE	R^2	MAPE	MEDAE	EVS
SIS	XGBoost	2.02 ± 1.02	6.17 ± 5.16	2.26 ± 1.07	-5.73 ± 13.13	0.09 ± 0.06	1.98 ± 1.10	0.62 ± 0.08
	LightGBM	1.44 ± 1.54	7.18 ± 9.42	1.93 ± 1.91	-2.16 ± 5.08	0.07 ± 0.08	1.47 ± 1.84	-0.20 ± 1.65
	DecisionTree	0.50 ± 0.70	1.11 ± 1.77	0.68 ± 0.83	-0.02 ± 2.28	0.03 ± 0.04	0.39 ± 0.70	0.42 ± 1.58
	SVR	3.47 ± 1.60	17.94 ± 14.47	3.92 ± 1.66	-17.98 ± 34.22	0.16 ± 0.09	3.42 ± 1.74	-1.89 ± 3.06
OKS	XGBoost	5.23 ± 3.40	41.66 ± 42.66	5.61 ± 3.27	-6.38 ± 10.48	0.19 ± 0.12	5.02 ± 3.50	0.56 ± 0.21
	LightGBM	7.03 ± 5.28	119.08 ± 128.43	9.09 ± 6.22	-47.70 ± 89.56	0.26 ± 0.23	6.67 ± 7.49	-14.68 ± 24.85
	DecisionTree	1.54 ± 2.29	9.48 ± 26.06	1.97 ± 2.44	-0.27 ± 3.61	0.05 ± 0.05	1.36 ± 2.48	0.69 ± 0.33
	SVR	8.21 ± 5.48	103.11 ± 115.27	8.77 ± 5.27	-16.74 ± 23.76	0.29 ± 0.17	8.15 ± 5.79	-0.12 ± 0.20

3.1 Results on the OKS Target Variable

Figure 1 visually summarizes the distribution of the regression evaluation metrics obtained across the nested LOPO folds for the OKS prediction task. Beyond the average values reported in Table 2, the boxplots highlight the substantial variability characterizing the evaluated models under strict participant-aware validation. In particular, the broad interquartile ranges and the presence of several extreme outliers observed for boosting-based and kernel-based approaches indicate unstable behavior across unseen participants, suggesting limited cross-subject generalization capability. Mainly, the consistently negative R^2 values observed across most experiments indicate that the relationship between the multimodal sensor features and the clinical OKS scores exhibits limited transferability across participants under strict cross-subject validation. These findings suggest the presence of substantial inter-subject variability and participant-specific behavioral patterns that cannot be easily transferred across individuals.

As illustrated in Fig. 1, the DT model exhibited the most compact distributions and the smallest number of extreme failures across nearly all evaluation metrics. The model achieved the lowest error values overall, including MAE, RMSE, MAPE, and MEDAE (Table 2). However, rather than indicating strong predictive capability, these results likely reflect the ability of simpler tree-based structures to avoid overfitting noisy and highly heterogeneous patterns present in the dataset. In contrast, more sophisticated approaches such as LightGBM, XGBoost, and SVR exhibited significantly larger variability and substantially worse R^2 scores, suggesting greater sensitivity to participant-specific noise and limited robustness under strict LOPO evaluation.

The instability of the boosting-based approaches was particularly evident for LightGBM, which produced extremely negative R^2 values and several large prediction outliers, as visible in Fig. 1. Similarly, SVR demonstrated poor generalization performance, characterized by large prediction errors and strongly negative explained variance values across multiple folds. These findings indicate that more complex nonlinear models may struggle to learn transferable representations from the current dataset, likely due to the limited cohort size and the high variability of recovery trajectories across patients.

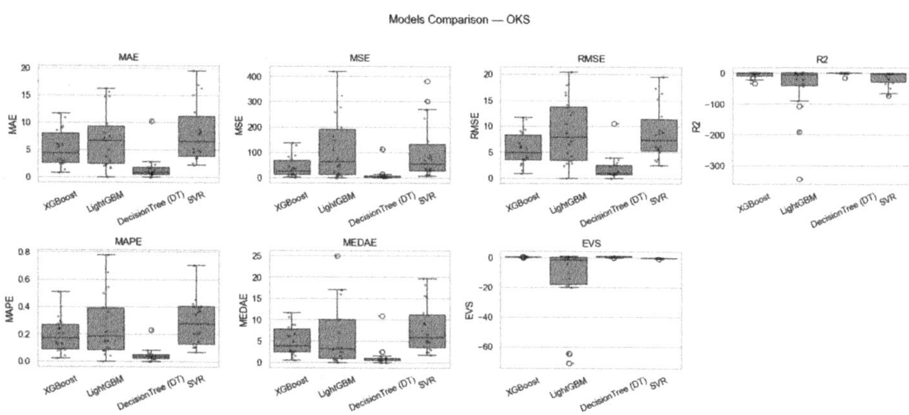

Fig. 1. Boxplots of the regression evaluation metrics for the OKS target variable obtained by the analyzed models.

3.2 Results on the SIS Target Variable

Figure 2 confirms similar trends for the SIS prediction task. Although some models achieved moderate prediction errors in selected folds, the overall distributions reveal weak and inconsistent generalization capability across participants. In particular, the large dispersion observed for LightGBM and SVR across MAE, RMSE, and R^2 indicates that these methods frequently fail to learn transferable behavioral representations from the multimodal sensing features.

The results obtained on the SIS prediction task confirm the trends observed for the OKS analysis and further highlight the difficulty of generalizing multimodal behavioral models across unseen participants. Despite moderate differences in prediction errors among the evaluated approaches, all models exhibited limited explanatory power under the strict nested LOPO evaluation framework, as summarized in Table 2.

The predominance of negative R^2 and EVS values across multiple folds indicates that the sensor-derived features maintain only weak associations with the SIS scores when evaluated in a participant-independent setting. This behavior suggests that social isolation dynamics may be strongly influenced by highly individualized behavioral and contextual factors that are difficult to capture through generalized machine learning models trained on a limited cohort.

As observed for OKS and illustrated in Fig. 2, the DT model achieved the lowest prediction errors and the most stable behavior across repeated experiments. In particular, the model obtained the best overall MAE, RMSE, MAPE, and MEDAE values reported in Table 2. However, these results should not be interpreted as evidence of highly accurate prediction, but rather as an indication that simpler tree-based approaches may be less prone to overfitting the participant-specific noise present in the dataset. In contrast, more complex models such as SVR and LightGBM displayed larger variability, higher error distributions, and substantially worse goodness-of-fit metrics.

Boosting-based approaches again demonstrated unstable behavior. Although XGBoost occasionally achieved competitive error values, the corresponding R^2 distributions remained highly variable and frequently negative, suggesting limited generalization capability across subjects. LightGBM exhibited the highest instability overall, with large interquartile ranges and several extreme outliers visible in Fig. 2, indicating sensitivity to the small-sample and high-noise characteristics of the dataset.

Collectively, these findings reinforce the hypothesis that participant heterogeneity constitutes the dominant challenge in multimodal passive monitoring of older adults recovering from LLF. The limited predictive performance observed across all methods suggests that current multimodal representations remain insufficient to fully capture the complexity of social and functional recovery trajectories under realistic cross-subject evaluation conditions.

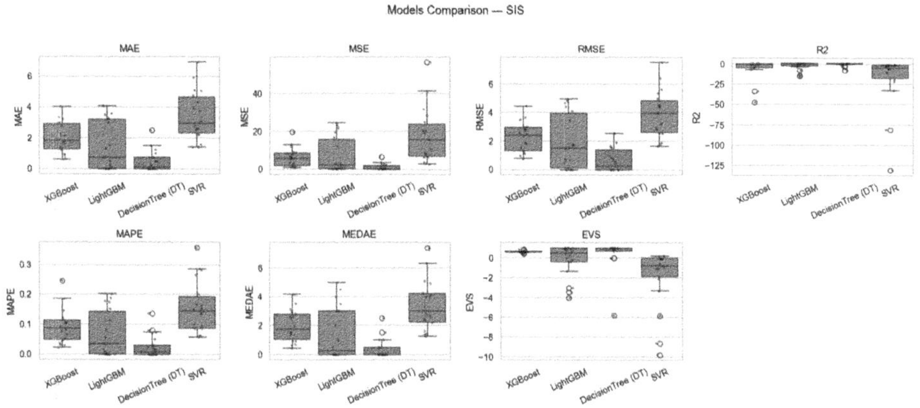

Fig. 2. Boxplots of the regression evaluation metrics for the SIS target variable obtained by the analyzed models.

4 Discussion and Conclusion

The experiments reveal a consistent and scientifically important trend: predicting clinical recovery outcomes from passive multimodal sensing data under strict participant-aware validation remains highly challenging. Although the DT approach systematically achieved the lowest prediction errors and the highest stability across repeated evaluations, the overall results indicate limited cross-subject generalization capability for all evaluated models.

In particular, the predominance of negative R^2 values across both OKS and SIS prediction tasks suggests that the relationship between multimodal behavioral features and clinical recovery scores exhibits limited transferability across participants. These findings highlight the substantial inter-subject variability characterizing older adults recovering from LLF and demonstrate the difficulty of learning generalized behavioral representations from passive sensing data in realistic deployment settings.

Interestingly, simpler models such as DTs demonstrated superior robustness compared to more sophisticated boosting-based and kernel-based approaches. Rather than indicating strong predictive capability, the improved stability of DT likely reflects its reduced tendency to overfit noisy or participant-specific patterns present in the dataset. In contrast, more complex methods such as LightGBM and SVR frequently produced unstable behavior, large performance variability, and strongly negative R^2 values across the nested LOPO folds. These observations suggest that highly parameterized models may struggle to learn transferable recovery representations under limited-data and high-heterogeneity conditions.

From a methodological perspective, these findings emphasize the importance of rigorous participant-aware evaluation protocols in multimodal digital health research. Standard random cross-validation strategies could substantially overestimate predictive performance by introducing participant-level leakage between training and testing samples. By adopting a strict nested LOPO framework, this study provides a more realistic estimate of how machine learning models generalize to previously unseen patients in real-world clinical scenarios.

The obtained results should also be interpreted considering several important limitations of the current dataset. Although the extended MAISON-LLF cohort represents a significant improvement over previous versions, the dataset still contains only 18 participants, with each individual contributing 56 daily observations collected over eight weeks. While sufficient for evaluating traditional tabular learning approaches, this scale remains relatively limited for reliably modeling complex temporal dependencies and participant-specific longitudinal dynamics. Although advanced tabular and sequence-based deep learning approaches may offer promising alternatives, including recent tabular foundation models such as TabPFN [6,13], their systematic evaluation was beyond the scope of this preliminary study. Future work will investigate whether these approaches can improve cross-subject generalization as the MAISON-LLF cohort continues to expand.

Another important consideration concerns the formulation of the prediction task itself. In this study, SIS and OKS were modeled as continuous regression targets. However, both clinical scores possess an inherent ordinal structure, where score ranges correspond to progressively different levels of social isolation and functional recovery. Future work may therefore investigate ordinal classification frameworks [5,12] capable of better preserving the clinical semantics of score progression and recovery severity levels.

Future research should focus on expanding the cohort size to enable more reliable modeling of temporal recovery trajectories and participant-specific behavioral dynamics. Larger longitudinal datasets would support the development of advanced ordinal, sequence-based, and personalized learning architectures capable of jointly modeling temporal dependencies and cross-subject variability. Additionally, hybrid global-local modeling strategies may provide a promising direction for improving generalization performance while preserving patient-specific adaptation.

Overall, this study demonstrates that rigorous participant-aware evaluation substantially changes the interpretation of predictive performance in multimodal digital health monitoring. Rather than highlighting raw predictive accuracy alone, the findings emphasize that cross-patient generalization remains a major open challenge in passive sensing applications for older adults recovering from LLF. These results motivate the need for larger longitudinal datasets, more robust personalized learning strategies, and evaluation protocols explicitly designed to account for inter-subject heterogeneity in real-world clinical AI systems.

References

1. Abedi, A., Chu, C.H., Khan, S.S.: Multimodal sensor dataset for monitoring older adults post lower limb fractures in community settings. Sci. Data **12**(1), 733 (2025)
2. Abedi, A., Colella, T.J., Pakosh, M., Khan, S.S.: Artificial intelligence-driven virtual rehabilitation for people living in the community: a scoping review. NPJ Dig. Med. **7**(1), 25 (2024)
3. Ahmad, A., Khan, S.: One-class classification with node embedding type features. Inf. Sci. **714**, 122204 (2025)
4. Dyer, S.M., et al.: A critical review of the long-term disability outcomes following hip fracture. BMC Geriat. **16**(1), 158 (2016)
5. Gómez-Orellana, A.M., et al.: Ordinal evolutionary artificial neural networks for predicting diabetic nephropathy progression. Appl. Soft Comput. **197**, 115153 (2026)
6. Hollmann, N., Müller, S., Eggensperger, K., Hutter, F.: Tabpfn: a transformer that solves small tabular classification problems in a second. arXiv preprint (2022)
7. Karagoz, B., Ashraf, A., Khan, S.S.: Supervised sequential contrastive regression: improving performance on imbalanced rehabilitation exercises. Biomed. Signal Process. Control **111**, 108347 (2026)
8. Khan, S.S., Gu, T., Spinelli, L., Wang, R.H.: Sensor-based assessment of social isolation in community-dwelling older adults: a scoping review. Biomed. Eng. Online **22**(1), 18 (2023)
9. Liciotti, D., Bernardini, M., Romeo, L., Frontoni, E.: A sequential deep learning application for recognising human activities in smart homes. Neurocomputing **396**, 501–513 (2020)
10. Paganelli, A.I., et al.: Real-time data analysis in health monitoring systems: a comprehensive systematic literature review. J. Biomed. Inf. **127**, 104009 (2022)
11. Párraga Vico, M.M., Morcillo Martínez, J.M., Gaitán-Guerrero, J.F., Herreros Bódalo, J.L., Espinilla Estévez, M., Cuevas Martínez, J.C.: Sensor-based technologies for the detection of unwanted loneliness in older adults: a systematic review. Sensors **26**(7), 2028 (2026)
12. Rosati, R., Romeo, L., Vargas, V.M., Gutierrez, P.A., Frontoni, E., Hervas-Martinez, C.: Learning ordinal-hierarchical constraints for deep learning classifiers. IEEE Trans. Neural Netw. Learn. Syst. (2024)
13. Sui, Y., Zhou, M., Zhou, M., Han, S., Zhang, D.: Table meets llm: can large language models understand structured table data? A benchmark and empirical study. In: Proceedings of the 17th ACM International Conference on Web Search and Data Mining, pp. 645–654 (2024)

Rolling-Window Feature Enrichment and Delta-Target Analysis in Cross-Participant Clinical Outcome Prediction

Mahmoud Z. A. Nijim, Shahrooz Abghari(✉), and Veselka Boeva

Department of Computer Science, Blekinge Institute of Technology, Karlskrona, Sweden
mani25@student.bth.se, {sab,vbx}@bth.se

Abstract. This study explores clinical outcome prediction using the MAISON-LLF dataset, which comprises multimodal sensor data collected from older adults recovering from lower-limb fractures in community settings. The study proposes a leakage-safe supervised regression pipeline for strict unseen-participant evaluation under Leave-One-Participant-Out cross-validation. The proposed approach uses rolling-window enrichment to convert daily sensor readings into short- and medium-term temporal summaries. It also evaluates an absolute-versus-delta target formulation, comparing absolute clinical-score prediction with change-from-baseline prediction using each participant's first available clinical assessment as the baseline. ElasticNet, LightGBM, and Random Forest are compared against a Naive Mean baseline. In absolute-score prediction, LightGBM achieves the lowest MAE for the Social Isolation Scale, while ElasticNet achieves the lowest MAE in both the Oxford Hip Score and the Oxford Knee Score. However, the improvements over the Naive Mean baseline are modest. The delta-target formulation substantially reduces MAE across targets, but the Naive Mean Delta baseline remains strong. The findings overall suggest that participant mean-shift strongly affects cross-participant prediction in MAISON-LLF data, and that simple baselines are necessary for interpreting model performance in small wearable-sensor rehabilitation datasets.

Keywords: Machine learning · Delta targets · Model personalization · Rolling-window · Assisted living · Sensor data

1 Introduction

Recent advances in Artificial Intelligence (AI), wearable sensing, and digital health technologies are changing how patient recovery can be monitored beyond

This study was submitted to the MAISON-LLF Data Challenge as part of the 9th ARIAL Workshop that took place alongside IJCAI-ECAI 2026.

S. S. Khan et al. (Eds.): IJCAI 2026, CCIS 3127, pp. 14–27, 2027.
https://doi.org/10.1007/978-981-92-5190-2_2

traditional clinical visits. AI-based methods can analyze complex health data and support prediction, decision-making, and personalized care when they are developed and validated responsibly [3]. At the same time, wearable and ambient sensors make it possible to continuously collect information about patients' daily activity patterns, sleep quality, and physiological signs in real-world settings. This is especially relevant for rehabilitation, where recovery is gradual and may not be fully captured during occasional clinic appointments. Sensor-based monitoring does not replace clinical assessment, but it can complement it by providing a more continuous view of recovery and by helping identify changes that may require earlier clinical attention.

A fracture to the lower limb places a substantial burden on patients and their caregivers, as well as on healthcare systems. Following surgery or conservative treatment, patients undergo months of rehabilitation with limited in-clinic monitoring. Wearable and ambient sensors can offer a practical solution for monitoring health recovery by continuously collecting information about movement, sleep patterns, heart rate, and activity levels. This information could be used to estimate functional recovery and social well-being between clinical assessments. However, these data are typically noisy, sparse, and highly participant-dependent. Furthermore, translating raw sensor signals into actionable clinical predictions requires Machine Learning (ML) models that can generalize successfully to new individuals. For this reason, data enrichment is often required prior to building ML models.

In this paper, we propose two strategies for enriching the MAISON-LLF data [2]: (i) rolling-window enrichment, which converts daily sensor readings into short- and medium-term temporal summaries so that the ML models are not limited to single-day measurements, and (ii) delta-target analysis, where absolute-score prediction is compared with change-from-baseline prediction. The latter strategy is motivated by clinical and longitudinal studies in which comparing follow-up values with baseline measurements is a common way to study change over time [13]. Therefore, our goal is not to introduce a cold-start deployable model, but to assess whether participant mean-shift is a major source of error under a Leave-One-Participant-Out Cross-Validation (LOPO-CV) setup.

The main contributions of this study are:

- a leakage-safe rolling-window enrichment strategy that converts daily sensor readings into short- and medium-term temporal summaries for strict evaluation of unseen participants using the LOPO-CV setup, and
- an absolute-versus-delta target analysis to examine whether participant mean-shift is a major source of error under LOPO-CV.

2 Related Works

The MAISON-LLF dataset was introduced by Abedi et al. [2] for monitoring older adults after lower-limb fractures in community settings. Their work established the value of collecting multimodal sensor data together with clinical outcome measures for studying recovery after fracture. The present study uses MAISON-LLF v4, which extends the setting to 18 participants and 1 008 daily records. While the dataset motivates clinical outcome prediction from multimodal sensor data, the small cohort size makes strict cross-participant evaluation difficult.

Previous work on MAISON-LLF includes the study by Boeva et al. [4], which used a multi-view multi-instance learning approach for health recovery monitoring in older adults. Their approach treats MAISON-LLF data as a multimodal learning problem, where information can come from different views and the clinical labels are not treated as simple day-level regression targets. This differs from the present study, which frames the task as supervised per-day regression of SIS (Social Isolation Scale), OHS (Oxford Hip Score), and OKS (Oxford Knee Score).

From a learning-approach perspective, the published MAISON-LLF work [4] can therefore be seen to focus mainly on data interpretation by treating the problem as a multimodal, multi-instance learning health monitoring task. The performance of this solution has been compared with the baselines published in [2] in two evaluation settings: 5-fold cross-validation and Leave-One-Sample-Out Cross-Validation (LOSO-CV). The authors have not used LOPO-CV in their experimental setup due to the unfavorable results reported in [2] for this type of validation. In contrast, this study focuses on strict unseen-participant prediction using LOPO-CV. This evaluation setting is important as random day-level splitting may overestimate performance when records from the same participant appear in both the train and test sets.

The gap addressed in this paper is therefore the combination of leakage-safe rolling-window enrichment, strict unseen-participant evaluation, and absolute-versus-delta target analysis on MAISON-LLF v4. While existing work [4] highlights the potential of multimodal sensing for recovery monitoring, the challenge of achieving participant-level generalization remains. This study addresses this issue by investigating whether enriched daily sensor features can outperform simple baselines, and whether change-from-baseline prediction can mitigate the mean-shift problem that hinders absolute-score prediction.

3 Methodology

The methodology is organized around a leakage-safe supervised learning workflow for evaluating MAISON-LLF v4 under strict cross-participant generalization. The workflow first prepares the daily multimodal records by removing leakage-prone variables, then enriches the remaining sensor features using rolling-window summaries to represent short- and medium-term recovery patterns. The enriched data are processed within the cross-validation pipeline, where data imputation, scaling, encoding, feature construction, model training, and hyperparameter tuning are fitted only on the training participants. Finally, the study evaluates two target formulations: absolute clinical-score prediction and delta-target analysis, in which scores are expressed as changes from a participant-specific baseline. This comparison examines whether participant-level mean-shift is a major source of error under LOPO-CV.

3.1 Data

MAISON-LLF v4 [1] is an extended version of the MAISON-LLF dataset described in [2]. The dataset contains 1 008 daily records from 18 participants, each observed for 56 days, spanning 84 raw columns. After removing identifiers and leakage-prone columns (Sect. 3.2), 46 safe daily sensor features remain across six modalities: acceleration (16), heart rate (6), motion (5), position (3), sleep (11), and step (5). Seven static demographic attributes are also retained per participant: sex, age, fracture type, relationship status, education, work status, and ethnicity.

The three regression targets are the Social Isolation Scale (SIS), Oxford Hip Score (OHS), and Oxford Knee Score (OKS). SIS captures social interaction and social isolation in older adults [11]. OHS and OKS are standard patient-reported clinical tools for assessing lower-limb physical function and pain [14,15].

Table 1 summarizes the score ranges and descriptive statistics of the three clinical targets used in this study. One of the dataset's defining properties is the significant variance in baseline scores between participants, which directly affects LOPO-CV behavior.

Table 1. Clinical outcome targets statistics of the MAISON-LLF used in this study.

Abbrev.	Full name	Range	Mean	Std. dev.
SIS	Social Isolation Scale	6–30	23.86	3.84
OHS	Oxford Hip Score	0–48	28.17	8.06
OKS	Oxford Knee Score	0–48	30.22	9.69

Preprocessing. A `ColumnTransformer` from scikit-learn [12] with two branches is applied inside each fold. The numeric branch applies median imputation followed by `RobustScaler`, which is preferred over standard scaling because accelerometry and step-count distributions are heavy-tailed. The categorical branch applies imputation followed by an ordinal encoder with an unknown-value sentinel for demographics. All transformers are fitted on training participants only and applied to the held-out participant.

3.2 The Proposed MAISON-LLF Data Enrichment Strategies

The proposed solution is built around two main strategies for data enrichment: *leakage-safe rolling-window enrichment* and *absolute-versus-delta target analysis*. The rolling-window enrichment strategy is used to represent short- and medium-term recovery patterns from daily sensor data. The delta-target analysis is then used to examine whether participant-level mean-shift is a major barrier to cross-participant generalization under LOPO-CV. These two data enrichment strategies are tested using three regression models: ElasticNet, LightGBM, and Random Forest. The models are used as evaluation tools rather than being the main contribution themselves.

Leakage-Safe Rolling-Window Enrichment: This study treats rolling-window enrichment as a key component of the proposed approach, rather than a minor preprocessing step. Clinical targets are assessed approximately every 14 days, whereas sensor data are available daily. Consequently, single-day sensor values may fail to capture short- and medium-term recovery patterns. Rolling windows are employed to convert daily sensor readings into temporal summaries that better reflect behavioral change throughout the rehabilitation period. These steps are explained in detail below:

(i) *Leakage removal.* Thirty-three columns are removed unconditionally before any feature engineering: the six SIS sub-item scores, the twelve OHS sub-item scores, and the twelve OKS sub-item scores, all of which are exact algebraic components of their respective targets; the concurrent clinical measurements `tug` and `chairstand`; and `clinical-timestamp`, which encodes the future assessment date. Each total score (SIS, OHS, OKS) is retained as a prediction target, while its sub-items are dropped as features. Cross-target exclusion is enforced inside the fold-specific `ColumnTransformer`, so that target columns never appear as features for any other target.

(ii) *Rolling-window enrichment.* Single-day sensor values capture only a snapshot, whereas clinical recovery unfolds over days and weeks. For each of the 46 safe sensor columns and for each 3-, 7-, and 14-day window, five statistics are computed per participant, sorted by timestamp: mean, standard deviation, minimum, maximum, and Ordinary Least Squares (OLS) slope against the integer time index. This produces $46 \times 3 \times 5 = 690$ rolling features. Combined with the 46 raw daily features and 7 demographic features, the final design matrix has 743 columns. All rolling computations are performed strictly within each participant's own time series. During LOPO-CV evaluation, the transformation is applied separately to the training participants and to the held-out participant, so no participant contributes sensor history to another participant's features. This enrichment step is important because it allows the models to use temporal recovery patterns rather than relying only on isolated daily sensor snapshots.

Absolute-Versus-Delta Target Analysis: This analysis compares two ways of defining the prediction target. In the absolute-score setting, the model directly predicts the original clinical score, e.g., SIS, OHS, or OKS. In the delta-target setting, the model instead predicts how much a participant's score changes from their own baseline. The baseline is defined as the participant's first available clinical assessment value. Change-from-baseline analysis is common when analyzing baseline and follow-up measurements in clinical studies [13].

A *Naive Mean baseline* is used as the simplest reference model in both the absolute-score and delta-target settings. This baseline does not use sensor features, demographic variables, or rolling-window features. Instead, for each outer LOPO-CV fold, it calculates the mean target value from the training participants only and uses this single value as the prediction for every record of the held-out participant. In the absolute-score setting, the mean is calculated from the original SIS, OHS, or OKS scores. In the delta-target setting, targets are first converted to change-from-baseline values, and the baseline then predicts the mean delta observed in the training participants. This provides a leakage-safe comparison point for assessing whether trained models add predictive value beyond a simple average-based prediction.

The delta target, y_{Δ}, is defined as $y_{\Delta} = y - y_{\text{baseline}}$, where y is the current observed clinical score for a participant, and y_{baseline} is the first available clinical score for the same participant. A delta value close to zero means that the participant's recovery is not showing significant progress. Evidently, a positive delta value indicates improvement in the recovery process, while a negative delta value indicates deterioration in the patient's recovery.

This transformation is used because participants may start from different baseline levels. In the absolute-score setting, the model has to predict both the participant's general score level and their change over time. In the delta-target setting, the participant-specific baseline is removed, so the prediction task focuses more directly on change during recovery. This allows the study to examine whether participant-level mean-shift is a major source of error under LOPO-CV.

Finally, note that this analysis is not considered a cold-start, deployable model because it requires at least one prior clinical assessment to define the participant-specific baseline.

4 Experiments and Implementation

A set of experiments was designed to evaluate the proposed enriched prediction pipeline. The models were trained using the rolling-window-enriched feature set, which represents short- and medium-term summaries of daily sensor data. The absolute-versus-delta analysis tested whether changing the target from an absolute clinical score to a change-from-baseline score reduces the effect of participant mean-shift. All reported model comparisons use the same leakage-safe LOPO-CV structure, so performance reflects generalization to unseen participants.

4.1 Experiment Design

The main evaluation uses LOPO-CV. The participant identifier is used as the grouping variable, yielding 18 outer folds. Each fold trains on 17 participants and evaluates on the remaining unseen participant. Since each participant contributes 56 daily records, each fold trains on approximately 952 rows and tests on 56 rows. This evaluation reflects the intended use case: predicting clinical outcomes for a participant not observed during training.

Hyperparameters are selected using nested cross-validation on the training participants only. Inside each outer LOPO fold, a 5-fold `GroupKFold` is used with participant ID as the grouping variable. This prevents any participant from appearing in both the inner training and validation splits. Grid search is scored using negative mean absolute error, and the best configuration is refit on all training participants before evaluation on the held-out participant.

For the absolute-score experiment, each model is trained to predict the original SIS, OHS, and OKS scores. For the delta-target experiment, the same LOPO-CV structure is used after transforming the targets into change-from-baseline values. The participant-specific baseline is defined as the first available clinical assessment value for that participant. The `clinical-timestamp` column is used only to identify the temporal order of assessments and is never used as a model feature. This makes the absolute and delta results directly comparable because the evaluation protocol remains the same and only the target definition changes.

All preprocessing steps, including imputation, scaling, and encoding, are applied within the fold-specific pipeline and fit only to the training participants. Models are trained only on the training participants. Rolling-window features are computed separately within each participant's own time series, so the held-out participant does not share sensor history with the training participants. The held-out participant is used only for the final evaluation in that fold.

4.2 Regression Models

The trainable regression models used in this study are ElasticNet, LightGBM, and Random Forest. These models were used to evaluate whether the rolling-window-enriched feature set provides predictive value under the leakage-safe LOPO-CV protocol. The Naive Mean baseline is used only as a reference point and does not use sensor, demographic, or rolling-window features.

ElasticNet [16] is used as the main sparse linear model. After rolling-window enrichment, the feature space contains many correlated temporal summaries. ElasticNet is suitable for this setting because the ℓ_1 component can shrink uninformative coefficients to zero, while the ℓ_2 component stabilizes estimation when predictors are correlated. Hyperparameters are tuned inside the inner cross-validation loop using $\alpha \in \{0.01, 0.1, 1, 10\}$, and ℓ_1-ratio values $\in \{0.1, 0.5, 0.9\}$.

LightGBM [9] is used as the main nonlinear model. It can capture nonlinear relationships and interactions between sensor modalities, while also supporting post-hoc interpretation through SHapley Additive exPlanations (SHAP) [10]. The tuned hyperparameters are selected from `n_estimators` $\in \{100, 300\}$, `learning_rate` $\in \{0.05, 0.1\}$, and `num_leaves` $\in \{31, 63\}$.

Random Forest [5] is retained as a secondary tree-based comparison because it provides a useful non-boosted reference model. It is not the central model in this study, but it is included to compare LightGBM with a non-boosted tree ensemble.

In addition to the main models mentioned above, several exploratory models were tested during development, including a demographics-only Ridge baseline, an untuned Ridge, and a tuned Ridge. These models were used to study whether simpler linear alternatives could outperform the Naive Mean baseline. As these models did not outperform the stronger working models, they are not central to the paper. Tuned Ridge is included only as an exploratory reference in the absolute-versus-delta comparison.

4.3 Evaluation Measures

Mean Absolute Error (MAE) is used as the primary evaluation metric. MAE is reported in the same units as the clinical outcome scores, which makes it easier to interpret the prediction errors for SIS, OHS, and OKS. Lower MAE indicates better predictive performance. The coefficient of determination, R^2, is also reported as a secondary metric for completeness. Negative R^2 values are expected in this setting because LOPO-CV evaluates each model on a participant whose baseline score may differ substantially from the training participants. In such cases, even moderate absolute prediction errors can lead to strongly negative R^2 values. Therefore, MAE is treated as the main metric when comparing models, while R^2 is used mainly to show the difficulty of cross-participant generalization.

SHAP [10] is used to interpret the LightGBM model. Global feature importance is computed from SHAP values for a representative outer fold, and features are also grouped by sensor modality to summarize which sensor types contribute most to each target. These explanations are treated as exploratory because feature importance may vary across held-out participants and has not been formally tested for fold-to-fold stability.

5 Results and Discussion

The results are interpreted in relation to two main questions: (i) whether the rolling-window-enriched sensor models improve over the Naive Mean baseline under LOPO-CV, and (ii) whether the delta-target formulation reduces the participant mean-shift problem. Table 2 summarizes the absolute-score prediction results. Table 3 and Fig. 1 compare absolute-score prediction with delta-target prediction. The implementations and experimental results are available at the GitHub.

5.1 Leakage-Safe Rolling-Window Enrichment

Absolute-Score Prediction. Table 2 summarizes the absolute-score MAE results across 18 LOPO folds for the main reported models. These results evaluate the direct prediction of the original SIS, OHS, and OKS scores. The Naive Mean baseline is included as the simplest reference model. It predicts the training-fold mean score for every row of the held-out participant and does not consider any features.

Table 2 reports the mean ± standard deviation MAE for the four models being compared. Lower MAE values are better, and the best result for each target is shown in bold. The absolute-score results are target-dependent: LightGBM gives the lowest SIS MAE (3.16 ± 1.07), while ElasticNet gives the lowest OHS MAE (6.23±4.50) and OKS MAE (8.03±5.08). However, the gains over the Naive Mean baseline are small. In addition, Random Forest slightly improves over the Naive Mean baseline for SIS, but Random Forest and LightGBM perform worse than the Naive Mean baseline on OHS and OKS, while ElasticNet performs slightly worse than the Naive Mean baseline on SIS. Therefore, the absolute-score results should be interpreted as modest and exploratory rather than as clear evidence of strong sensor-driven prediction.

Table 2. The mean and standard deviation of the absolute-score LOPO-CV MAE results for predicting the SIS, OHS, and OKS are shown for the four studied models.

Model	SIS MAE	OHS MAE	OKS MAE
Naive Mean	3.29 ± 1.65	6.49 ± 4.81	8.32 ± 5.29
ElasticNet	3.45 ± 1.91	**6.23 ± 4.50**	**8.03 ± 5.08**
Random Forest	3.25 ± 1.27	8.73 ± 6.14	9.24 ± 4.82
LightGBM	**3.16 ± 1.07**	9.39 ± 6.20	9.40 ± 5.00

For SIS prediction, LightGBM performs best and is closely followed by Random Forest. Both are slightly below the Naive Mean baseline, while ElasticNet is slightly higher. The difference between LightGBM and the Naive Mean baseline is approximately 4% in mean MAE, which is small compared with the fold-level standard deviation. For OHS and OKS, ElasticNet achieves the best performance and marginally outperforms the Naive Mean baseline. Random Forest performance slightly improves over the Naive Mean baseline for SIS, but it does not improve over the baseline for OHS and OKS. LightGBM also performs worse than the Naive Mean baseline on OHS and OKS. This suggests that nonlinear modeling does not overcome the between-participant variability in this small-cohort LOPO setting.

Negative R^2 under LOPO-CV. All models, including the Naive Mean baseline, produce negative R^2 values under LOPO-CV. This is expected in this dataset because R^2 is computed relative to the held-out participant's own mean score. If the held-out participant has a baseline score that differs substantially from that of the training participants, even moderate absolute errors can lead to a strongly negative R^2. For this reason, a negative R^2 is interpreted as evidence of participant-level mean-shift and small-cohort difficulty, rather than a failure of the evaluation pipeline. MAE is therefore treated as the primary metric because it is easier to interpret in clinical score units.

5.2 Absolute Versus Delta-Target Prediction

Table 3 compares the original absolute-score prediction task with the delta-target formulation. In the delta setting, the target is transformed into changes from each participant's own baseline, so the model predicts y_{Δ} rather than the original score y. The Δ columns report the delta-target results. Tuned Ridge is included only as an exploratory reference. The Naive Mean Delta baseline follows the same idea as the ordinary Naive Mean baseline, but after the target has been transformed. It predicts the average change-from-baseline value observed in the training participants and applies that same value to every row of the held-out participant.

The delta formulation reduces MAE across all models and targets. This supports the interpretation that participant mean-shift is a major source of error in the absolute-score LOPO setting. However, the Naive Mean Delta baseline is also very strong. ElasticNet only marginally improves over the Naive Mean Delta baseline for SIS and OHS, and for OKS, the Naive Mean Delta baseline is slightly better than ElasticNet. This means that the delta-target analysis mainly shows that removing participant-specific baseline levels makes the task easier. However, it does not prove that the sensor models strongly learn within-participant change beyond simple baseline subtraction.

Table 3. The mean and standard deviation of the absolute-score versus delta-target LOPO-CV MAE results for predicting the SIS, OHS, and OKS are shown for the five studied models. Bold values indicate the lowest mean MAE based on the unrounded values.

Model	SIS MAE		OHS MAE		OKS MAE	
	Absolute	Δ	Absolute	Δ	Absolute	Δ
Naive Mean	3.29 ± 1.65	1.86 ± 1.14	6.49 ± 4.81	2.86 ± 1.90	8.32 ± 5.29	$\mathbf{3.32 \pm 2.10}$
Ridge (tuned)	5.04 ± 2.69	3.89 ± 2.91	13.06 ± 11.22	5.03 ± 2.18	13.62 ± 8.74	7.16 ± 7.78
ElasticNet	3.45 ± 1.91	$\mathbf{1.86 \pm 1.20}$	$\mathbf{6.23 \pm 4.50}$	$\mathbf{2.83 \pm 1.88}$	$\mathbf{8.03 \pm 5.08}$	3.35 ± 2.08
Random Forest	3.25 ± 1.27	2.27 ± 1.03	8.73 ± 6.14	3.91 ± 1.76	9.24 ± 4.82	3.63 ± 2.02
LightGBM	$\mathbf{3.16 \pm 1.07}$	2.55 ± 1.12	9.39 ± 6.20	3.65 ± 1.87	9.40 ± 5.00	3.54 ± 2.05

The strongest effect in Table 3 and Fig. 1 is not the difference between ML models, but the difference between absolute-score and delta-target prediction. For example, the Naive Mean baseline drops from 3.29 to 1.86 MAE for SIS, from 6.49 to 2.86 for OHS, and from 8.32 to 3.32 for OKS. This pattern indicates that much of the absolute-score error is caused by differences in participants' baseline levels. At the same time, the closeness of ElasticNet to the Naive Mean Delta baseline suggests that the remaining within-participant variation is difficult to learn from only 17 training participants per fold.

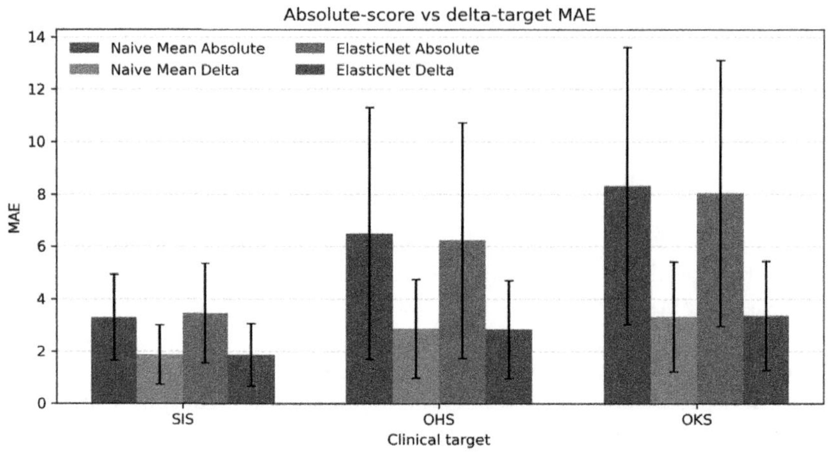

Fig. 1. A comparison of the absolute-score and delta-target LOPO-CV MAE for Naive Mean and ElasticNet. Delta targets reduce MAE, while the Naive Mean Delta baseline remains close to ElasticNet.

5.3 Exploratory SHAP Interpretation

SHAP analysis was used to inspect LightGBM feature importance in a representative held-out fold. Modality-level summaries suggest that (i) acceleration and step features are important for SIS prediction, (ii) sleep features are important for OHS prediction, and (iii) motion and sleep features are important for OKS prediction. These findings are exploratory because SHAP values were not formally assessed for fold-to-fold stability across all held-out participants. Therefore, the SHAP results should be interpreted as a qualitative indication of possible sensor-modality relevance rather than as stable evidence of clinical feature importance.

5.4 Statistical Considerations

The fold-level standard deviations are large relative to the mean MAE differences between models. Therefore, the Friedman test [6,7], a non-parametric test, is applied separately for each target to assess whether the model rankings differed consistently across the 18 LOPO folds. Hence, the five built models are compared in both the absolute-score and delta-target settings: Naive Mean, Ridge, ElasticNet, Random Forest, and LightGBM. We define the null hypothesis as the compared models having similar MAE distributions across the LOPO folds. A significant Friedman statistic indicates an overall difference among models, but it does not identify which specific models differ from each other. Upon rejecting the null hypothesis at the 0.05 significance level, a Nemenyi post-hoc test [6] is used to compare the average model ranks. The Nemenyi test calculates Critical

Difference (CD) as follows:

$$CD = q_\alpha \sqrt{\frac{k(k+1)}{6N}}, \quad (1)$$

where q_α with significance level $\alpha = 0.05$ and $k = 5$ is 2.728, leading to $CD = 1.438$ for $N = 18$ LOPO folds. Two models are considered significantly different when the difference between their average ranks exceeds the critical difference. A lower average rank indicates a lower MAE and, therefore, better performance.

Table 4. Friedman test statistics (χ^2) comparing model MAE values across the 18 LOPO folds. Rows with p-values < 0.05 (in **bold**) indicate a significant overall model difference.

Setting	Target	χ^2	p-value
Absolute-score	SIS	9.29	0.0543
Absolute-score	OHS	22.62	**0.0002**
Absolute-score	OKS	9.16	0.0573
Delta-target	SIS Δ	27.87	$<$ **0.0001**
Delta-target	OHS Δ	13.05	**0.0110**
Delta-target	OKS Δ	22.04	**0.0002**

The Friedman test results in Table 4 show that, in the absolute-score setting, only OHS has a significant overall model difference, while SIS and OKS do not. In the delta-target setting, the Friedman statistic indicates significant overall model differences across all three targets. Therefore, Nemenyi post-hoc comparisons are conducted only for the cases where the Friedman test p-value < 0.05.

In the analysis of absolute-score OHS, the Nemenyi post-hoc results indicate that both ElasticNet and Naive Mean achieved significantly lower average ranks compared to LightGBM and Ridge. In the delta-target settings, the significant differences vary depending on the target. For SIS Δ, both ElasticNet and Naive Mean achieve significantly lower average ranks than LightGBM and Ridge. For OHS Δ, Naive Mean and ElasticNet earn significantly lower average ranks than Ridge. Lastly, for OKS Δ, Ridge performs significantly worse than the rest.

The Nemenyi post-hoc results do not show a consistent statistically significant advantage of the supervised learning models over the Naive Mean baseline. The significant differences are target-dependent and mainly indicate that some models perform worse for specific targets, especially LightGBM and Ridge for absolute-score OHS, and Ridge for delta-target prediction. Overall, these results support a cautious interpretation: strict participant-level generalization remains difficult in this small-cohort LOPO-CV setting, and delta-target analysis suggests that participant mean-shift is an important source of prediction error.

6 Conclusion

This exploratory study evaluated two strategies: rolling-window and delta-target analysis for enriching MAISON-LLF data. The goal was to determine whether they could improve the generalization of the three built supervised models (Random Forest, ElasticNet, and LightGBM) and support personalized monitoring of health recovery.

The rolling-window enrichment strategy allowed the models to use short- and medium-term summaries of daily sensor data rather than isolated daily measurements. However, the absolute-score results suggest that enriched sensor features do not automatically lead to strong generalization for unseen participants. The delta-target analysis changed the prediction target from absolute clinical scores to change-from-baseline values, reducing MAE across all targets and models. This finding suggests that differences in baseline measurements between participants are a significant source of prediction error in Leave-One-Participant-Out Cross-Validation.

The overall experimental results show that strict cross-participant prediction remains difficult in this small-cohort setting. LightGBM achieved the lowest MAE for the Social Isolation Scale, while ElasticNet achieved the lowest MAE for both the Oxford Hip Score and the Oxford Knee Score. Random Forest slightly improved over the Naive Mean baseline for SIS, but did not improve over the baseline for OHS and OKS. However, the improvements over the Naive Mean baseline were modest, and model performance was target-dependent. The results of the Friedman and Nemenyi tests further support this cautious interpretation, since the supervised models did not show a consistent statistically significant advantage over the Naive Mean baseline.

Future work should compare rolling-window enrichment against non-enriched feature baselines. If larger MAISON-LLF cohorts become available, data-efficient tabular learning methods, such as TabPFN [8], could also be evaluated.

References

1. Abedi, A., Chu, C., Khan, S.S.: Maison-llf: multimodal sensor dataset for monitoring older adults post lower-limb fractures in community settings (2025). https://doi.org/10.5281/zenodo.14803293
2. Abedi, A., Chu, C.H., Khan, S.S.: Multimodal sensor dataset for monitoring older adults post lower limb fractures in community settings. Sci. Data **12**(1), 733 (2025)
3. Bajwa, J., Munir, U., Nori, A., Williams, B.: Artificial intelligence in healthcare: transforming the practice of medicine. Future Healthcare J. **8**(2), e188–e194 (2021). https://doi.org/10.7861/fhj.2021-0095
4. Boeva, V., Ojutkangas, A.J.O., Abghari, S.: Multi-view multi-instance learning for health recovery monitoring in older adults. In: Bruno, P., Calimeri, F., Cauteruccio, F., Dragoni, M., Stella, F., Terracina, G. (eds.) Artificial Intelligence for Healthcare, and Hybrid Models for Coupling Deductive and Inductive Reasoning, pp. 398–411. Springer, Cham (2026). https://doi.org/10.1007/978-3-032-16708-8_32
5. Breiman, L.: Random forests. Mach. Learn. **45**(1), 5–32 (2001)

6. Demšar, J.: Statistical comparisons of classifiers over multiple data sets. J. Mach. Learn. Res. **7**(1), 1–30 (2006). http://jmlr.org/papers/v7/demsar06a.html
7. Friedman, M.: The use of ranks to avoid the assumption of normality implicit in the analysis of variance. J. Am. Stat. Assoc. **32**(200), 675–701 (1937)
8. Hollmann, N., et al.: Accurate predictions on small data with a tabular foundation model. Nature **637**(8045), 319–326 (2025)
9. Ke, G., et al.: LightGBM: a highly efficient gradient boosting decision tree. Adv. Neural Inf. Process. Syst. (NeurIPS) **30** (2017)
10. Lundberg, S.M., Lee, S.I.: A unified approach to interpreting model predictions. Adv. Neural Inf. Process. Syst. (NeurIPS) **30** (2017)
11. Nicholson, N.R., Feinn, R., Casey, E.A., Dixon, J.: Psychometric evaluation of the social isolation scale in older adults. Gerontologist **60**, e491–e501 (2020). https://doi.org/10.1093/geront/gnz083
12. Pedregosa, F., et al.: Scikit-learn: machine learning in python. J. Mach. Learn. Res. **12**, 2825–2830 (2011)
13. Vickers, A.J., Altman, D.G.: Statistics notes: analysing controlled trials with baseline and follow up measurements. BMJ **323**(7321), 1123–1124 (2001). https://doi.org/10.1136/bmj.323.7321.1123
14. Whitehouse, S.L., Blom, A.W., Taylor, A.H., Pattison, G.T.R., Bannister, G.C.: The oxford knee score; problems and pitfalls. Knee **12**, 287–291 (2005). https://doi.org/10.1016/j.knee.2004.11.005
15. Wylde, V., Learmonth, I.D., Cavendish, V.J.: The oxford hip score: the patient's perspective. Health Qual. Life Outcomes **3**, 66 (2005). https://doi.org/10.1186/1477-7525-3-66
16. Zou, H., Hastie, T.: Regularization and variable selection via the elastic net. J. Royal Stat. Soc. Ser. B (Stat. Methodol.) **67**(2), 301–320 (2005)

Multimodal Forecasting of Psychosocial and Functional Recovery in Older Adults After Lower Limb Fracture

Jiyoo Noh[1(✉)], Wantong Huo[1], Kaung Myat Kyaw[2], Abigayle Brown[1], Khush Agarwal[2], and Jonathan H. Chan[2]

[1] University of Toronto, Toronto, Canada
{jiyoo.noh,vanessa.huo,abbey.brown}@mail.utoronto.ca
[2] King Mongkut's University of Technology Thonburi, Bangkok, Thailand
{kaungmyat.kyaw,khush.agar}@kmutt.ac.th, jonathan@sit.kmutt.ac.th

Abstract. Wearable and passive sensing technologies enable continuous monitoring of recovery after lower-limb fractures (LLF). However, existing approaches focus on concurrent outcome estimation rather than prospective prediction. We formulate rehabilitation monitoring as a temporal forecasting task using the MAISON-LLF dataset, predicting future Oxford Hip Score (OHS) and Social Isolation Score (SIS) assessments from prior multimodal sensor observations. We propose a multi-task GRU-based forecasting framework with SHAP-guided feature selection, learnable feature gates, masked sequence handling, and task-specific recency-aware attention. Experiments on 18 community-dwelling older adults with four biweekly assessments per participant show that joint OHS and SIS prediction improves generalization over independently trained single-task baselines, particularly for OHS total-score forecasting and SIS correlation-based performance. However, SIS prediction remains less well calibrated, reflecting the difficulty of forecasting psychosocial outcomes from a small heterogeneous cohort. Overall, the results demonstrate the feasibility of prospective functional and psychosocial recovery forecasting from longitudinal multimodal sensing data. Code is available at: https://github.com/IC2-Lab-KMUTT/maison-llf.

Keywords: Temporal forecasting · multimodal learning · wearable sensing · MAISON-LLF

1 Introduction

Lower-limb fractures (LLF) in older adults often result in prolonged recovery trajectories characterized by reduced mobility, impaired functional independence, and psychosocial challenges including increased social isolation [7]. Recovery after discharge is inherently dynamic, evolving over weeks through interactions among physical activity, sleep behaviour, mobility, and social engagement [4,19].

S. S. Khan et al. (Eds.): IJCAI 2026, CCIS 3127, pp. 28–44, 2027.
https://doi.org/10.1007/978-981-92-5190-2_3

Conventional follow-up practices rely on sparse clinical assessments and self-reported questionnaires, limiting the ability to continuously monitor recovery progression and identify early deterioration [1]. Wearable and passive sensing technologies offer an opportunity to capture behavioural and physiological information in real-world settings, enabling richer longitudinal representations of patient recovery [12,17].

The MAISON-LLF dataset [1] provides a multimodal longitudinal resource for studying LLF recovery in older adults. The dataset combines daily sensor-derived features from smartphones, smartwatches, sleep sensors, and in-home motion sensors with repeated clinical assessments collected during the post-discharge recovery period. Among the available outcomes, the Oxford Hip Score (OHS) [14] measures hip-related pain and physical function, while the Social Isolation Score (SIS) [15] measures social interaction and isolation in older adults, both acquired in biweekly assessments [6,15].

Prior analyses of MAISON-LLF demonstrated the feasibility of using machine learning to estimate clinical outcomes from multimodal sensor features [1]. However, the original validation formulation assigned clinical labels to daily, weekly, or biweekly sensor windows preceding each assessment, making the task closer to concurrent outcome estimation than prospective forecasting. In realistic deployment settings, future patient states must be predicted using only information available before the clinical assessment date. Improper handling of temporal structure in longitudinal data can introduce information leakage and lead to overly optimistic estimates of model performance [5,10]. Forecasting future recovery outcomes is clinically relevant because care decisions often depend on anticipating patient status before deterioration becomes evident. Predicting future OHS and SIS measurements from prior sensor histories could help identify patients at risk for delayed functional recovery or worsening psychosocial outcomes. However, this problem is challenging because clinical labels are sparse, sensor measurements are heterogeneous, and patient histories vary in length.

In this work, we use MAISON-LLF to formulate OHS and SIS prediction as a time-series forecasting task. For each patient and clinical assessment, cumulative sensor histories are restricted to observations available before the target assessment, with a one-week forecasting horizon. To handle the resulting variable-length, multimodal sequences, we propose a Gated Recurrent Unit-based (GRU-based) multi-task forecasting model incorporating three key components: (i) SHapley Additive exPlanations (SHAP) guided feature selection and learnable feature gating [13], (ii) target-specific recency-biased attention that adaptively weights recent observations more heavily [2], and (iii) masked sequence handling for variable-length patient histories. Through this approach, our work moves beyond same-window outcome estimation toward prospective prediction settings that better reflect real-world rehabilitation monitoring. By forecasting future recovery outcomes from prior multimodal sensor histories, the proposed framework supports the development of personalized monitoring systems for earlier intervention and adaptive rehabilitation.

The contributions of this work are summarized as follows:

1. We formulate OHS and SIS prediction using MAISON-LLF as a leakage-safe temporal forecasting task under strict chronological separation between inputs and targets.
2. We construct expanding-window patient histories to model recovery progression from variable-length multimodal sensor sequences.
3. We develop an attention-based recurrent forecasting model integrating SHAP-guided feature weighting, recency-aware temporal weighting, and masked sequence handling.

2 Related Work

2.1 Sensor-Based Monitoring After Lower Limb Fracture

Remote monitoring has become increasingly important for older adults recovering from LLF, as recovery is often associated with reduced mobility, prolonged functional limitations, and loss of independence [7]. Prior studies [1,11] have used wearable accelerometers, step counts, and in-home sensors to assess activity and rehabilitation progress. However, many existing approaches rely on short monitoring periods or single sensor modalities, limiting their ability to capture longitudinal recovery patterns in real-world home environments. The MAISON-LLF dataset addresses this limitation by providing continuous multimodal sensing from smartphones, smartwatches, motion sensors, and sleep-tracking mats during post-discharge recovery.

2.2 Social Isolation and Functional Decline Prediction

Social isolation and functional decline are clinically important outcomes that are difficult to monitor through routine care alone. Existing sensor-based studies [11, 16,17] have associated mobility, sleep, heart rate, and in-home activity patterns with social isolation or physical functioning. However, these works study the outcomes separately and focus primarily on correlation rather than predictive modeling. MAISON-LLF is distinctive because it includes both SIS and OHS assessments, enabling joint modeling of psychosocial and functional recovery outcomes [1].

2.3 Multimodal Learning with Sparse Clinical Labels

A key challenge in healthcare forecasting is that sensor data are collected continuously, while clinical outcomes are measured sparsely. In MAISON-LLF, daily sensor features are available throughout the recovery period, whereas clinical questionnaires are collected approximately every two weeks. The prior baseline approach [1] assigns the same clinical label to daily, weekly, or biweekly windows preceding each assessment. While useful for outcome estimation, such formulations do not fully represent prospective forecasting. In contrast, our work constructs expanding longitudinal histories and predicts future clinical scores using only data available before the target clinical timestamp.

2.4 Temporal Models for Longitudinal Healthcare Forecasting

Sequential models such as long short-term memory networks (LSTMs) [8], GRUs [3], temporal convolutional networks [9], and transformers [18] are well suited for longitudinal healthcare data because they can model temporal dependencies and recovery trajectories. Prior MAISON-LLF validation included sequential models evaluated on fixed-length windows. The proposed approach instead uses variable-length patient histories with masked sequence handling and target-specific decay attention, allowing the model to learn recovery dynamics as additional sensor history accumulates over time. To the best of our knowledge, this is the first work to formulate MAISON-LLF as a leakage-safe longitudinal forecasting task with strict chronological input-target separation and patient-level evaluation.

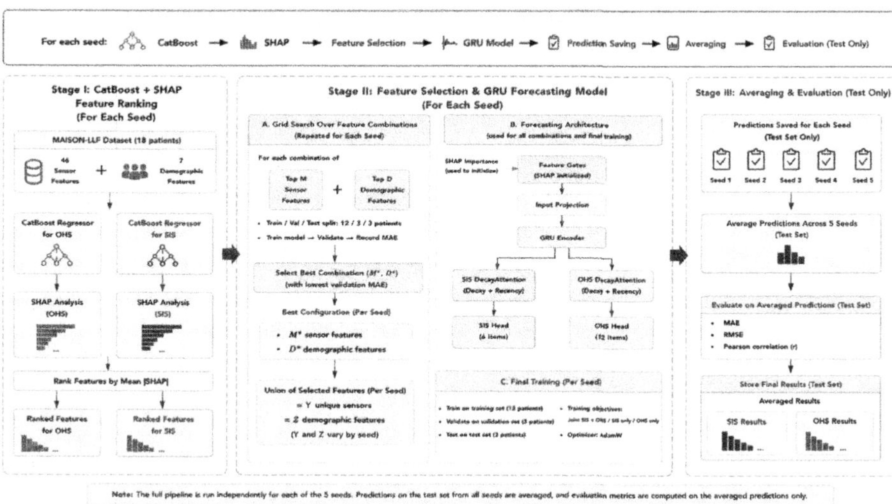

Fig. 1. Overview of the proposed forecasting pipeline. CatBoost and SHAP are used to rank sensor and demographic features for OHS and SIS prediction. A grid search identifies the optimal feature subset, which is used to train a GRU forecasting model with SHAP-initialized feature gates and attention. Performance is evaluated on a held-out test set using item-level and total-score metrics for both outcomes.

3 Methods

3.1 Problem Formulation

The full experiment pipeline is shown in Fig. 1. We formulate recovery prediction as a 7-day-ahead forecasting task. The objective is to predict both the item-level and total OHS and SIS outcomes on the next clinical assessment. Clinical assessments are conducted biweekly. Let n denote the clinical assessment index, and let T_n denote the day on which the nth assessment occurs. For each assessment, the model predicts the outcomes 7 days in advance using only observations collected up to day $T_n - 7$. For example, predicting scores at the first assessment ($T_1 = 14$) uses data from the first 7 days, while predicting scores at the second assessment ($T_2 = 28$) uses data from the first 21 days. For patient i, let $x_{i,d} \in \mathbb{R}^p$

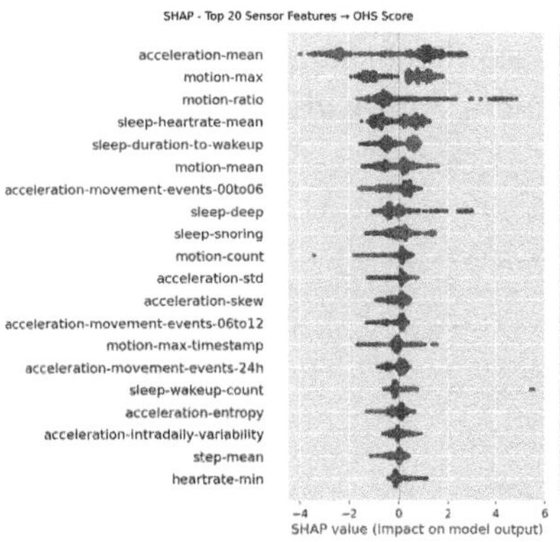

(a) SHAP summary plot showing the top 20 sensor features ranked by importance for predicting OHS score.

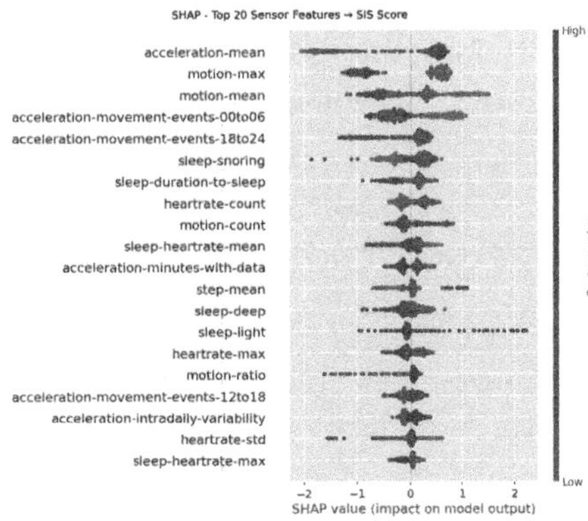

(b) SHAP summary plot showing the top 20 sensor features ranked by importance for predicting SIS score

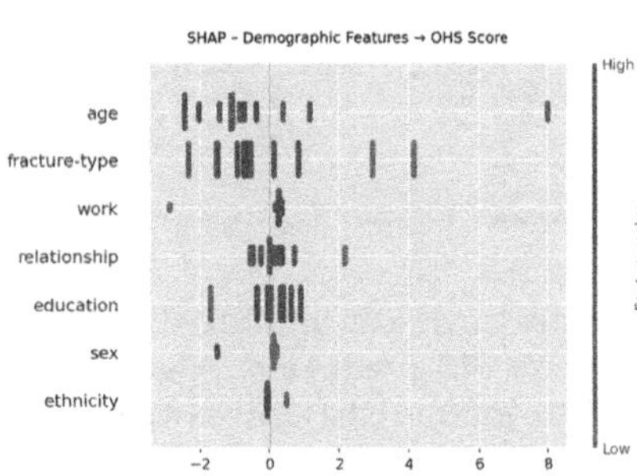

(c) SHAP summary plot showing the 7 demographic features ranked by importance for predicting OHS score

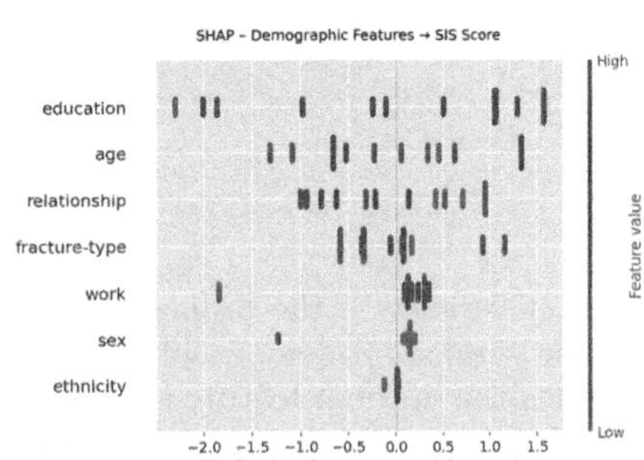

(d) SHAP summary plot showing the 7 demographic features ranked by importance for predicting SIS score

Fig. 2. SHAP summary plots showing sensor and demographic feature importance for predicting OHS and SIS scores.

denote the vector of p selected features recorded on day d. Here, p denotes the number of selected input features after feature selection. The input sequence used to predict the nth assessment is

$$X_{i,n} = [x_{i,1}, x_{i,2}, \ldots, x_{i,T_n-7}] \in \mathbb{R}^{(T_n-7)\times p}. \tag{1}$$

In that case, $L_{i,n}$ denotes the number of valid observations available for patient i when predicting assessment n. Variable-length sequences are handled using padding and masking during model training. The prediction task is formulated as

$$\hat{\mathbf{y}}_{i,n}^{(q)} = f_q(X_{i,n}), \tag{2}$$

where $q \in \{\text{SIS}, \text{OHS}\}$ denotes the target clinical outcome, $f_q(\cdot)$ denotes the forecasting model, and $\hat{\mathbf{y}}_{i,n}^{(q)}$ denotes the predicted questionnaire item scores for patient i at the nth assessment. Specifically, $\hat{\mathbf{y}}_{i,n}^{(\text{SIS})} \in \mathbb{R}^6$ and $\hat{\mathbf{y}}_{i,n}^{(\text{OHS})} \in \mathbb{R}^{12}$ correspond to the 6 SIS and 12 OHS questionnaire items, respectively. The total SIS and OHS scores are obtained by summing the predicted item-level scores.

3.2 CatBoost-Based Feature Selection

SHAP-based feature selection is used to retain only the most informative sensor and demographic variables for the subsequent forecasting model. This is done because not all available variables contribute equally to recovery prediction. We follow the feature selection procedure of the original dataset study [1]. Separate CatBoostRegressor models are trained for SIS and OHS using 46 sensor features and 7 demographic variables. Feature selection is performed using the training set only, following the same train/validation/test split used in the subsequent forecasting experiments. The ranked SHAP features of OHS and SIS are shown in Fig. 2. For each outcome $q \in \{\text{SIS}, \text{OHS}\}$, the importance of feature j is defined as the mean absolute SHAP value across all training observations:

$$I_j^{(q)} = \frac{1}{N} \sum_{r=1}^{N} \left| \phi_{r,j}^{(q)} \right|, \tag{3}$$

where $\phi_{r,j}^{(q)}$ is the SHAP value of feature j for training observation r, and N is the number of training observations. For each outcome, the top-M sensor features and top-D demographic features are selected. The union of the selected SIS and OHS features forms the shared input feature set used by the forecasting model. The number of features in this shared input set is denoted by p.

3.3 Feature Gates

Before training the model, the selected sensor features and selected static demographic features are combined into a single input vector at each time-step. Static demographic features are repeated across time so that each time-step contains both sensor and demographic information. Let $z_{i,n,\tau} \in \mathbb{R}^p$ denote the combined input vector at time-step τ for patient i when predicting assessment n. This input vector is passed through a learnable feature gate:

$$\tilde{z}_{i,n,\tau} = \mathbf{w}_g \odot z_{i,n,\tau}, \tag{4}$$

where $\mathbf{w}_g \in \mathbb{R}^p$ is a vector of learnable gate weights, and $\odot$ denotes element-wise multiplication. The gate weights are initialized using the SIS and OHS SHAP importance values. During training, they are constrained to be positive and normalized to have unit mean:

$$\mathbf{w}_g = \frac{\text{softplus}(\boldsymbol{\theta})}{\frac{1}{p}\sum_{j=1}^{p} \text{softplus}(\theta_j)}, \tag{5}$$

where $\boldsymbol{\theta} \in \mathbb{R}^p$ denotes the unconstrained learnable gate parameters. This gate allows the model to learn the relative importance of each input feature while avoiding global rescaling of the input sequence.

3.4 GRU-Based Forecasting

After feature gating, the input sequence is projected into a shared latent representation using a linear projection layer followed by layer normalization, GELU activation, and dropout. The projected sequence is then passed to a GRU encoder. For patient i and assessment n, the GRU produces one hidden representation for each valid time-step:

$$h_{i,n,1}, h_{i,n,2}, \ldots, h_{i,n,L_{i,n}} = \text{GRU}\left(\tilde{z}_{i,n,1}, \tilde{z}_{i,n,2}, \ldots, \tilde{z}_{i,n,L_{i,n}}\right), \tag{6}$$

where $L_{i,n}$ is the number of valid observations available for that patient-assessment pair. Variable-length sequences are handled using packed sequences, and padded time-steps are masked during attention. Rather than relying only on the final GRU hidden state, the model uses task-specific decay attention modules to aggregate information across the full sequence. Separate attention modules are used for SIS and OHS, allowing each outcome to learn its own temporal weighting pattern. The attention mechanism assigns larger weights to more informative time-steps while including a learnable recency bias that discounts older observations. For each task $q \in \{\text{SIS}, \text{OHS}\}$, the attention module produces a context vector:

$$c_{i,n}^{(q)} = \sum_{\tau=1}^{L_{i,n}} \alpha_{i,n,\tau}^{(q)} h_{i,n,\tau}, \tag{7}$$

where $\alpha_{i,n,\tau}^{(q)}$ is the attention weight assigned to time-step τ. Attention weights are computed only over valid time-steps; padded time-steps are masked out. Finally, task-specific prediction heads map the context vectors to item-level outcomes:

$$\hat{\mathbf{y}}_{i,n}^{(q)} = g_q(c_{i,n}^{(q)}), \tag{8}$$

where g_q denotes the prediction head for outcome q. The SIS prediction head outputs six item scores, and the OHS prediction head outputs twelve item scores.

3.5 Training and Evaluation

The model is trained end-to-end using AdamW on three training configurations: joint SIS and OHS prediction, SIS-only prediction, and OHS-only prediction. The training objective is the sum of the SIS and OHS item-level Smooth L1 losses. Training uses gradient clipping and early stopping based on validation loss. Performance is evaluated on held-out test patients using mean absolute error (MAE), root mean squared error (RMSE), coefficient of determination R^2, and Pearson correlation coefficient r. To assess robustness to random initialization, all experiments are repeated across five independent random seeds. Performance is reported for each seed separately to quantify variability across training runs. In addition, we construct a seed-level ensemble by averaging the item-level output predictions from the five independently trained models before computing total-score predictions. The ensemble results are then compared against the individual-seed results to determine whether averaging across seeds improves stability and predictive performance.

4 Experiments

4.1 Dataset and Experimental Setup

Experiments are conducted using the MAISON-LLF dataset, which contains time-series multimodal sensor measurements collected from 18 community-dwelling older adults recovering from LLF. Each patient underwent four biweekly clinical assessments, yielding 72 prediction instances in total. The clinical target variables (OHS, OKS, SIS, TUG, and chairstand) are excluded from the model input to avoid information leakage, and a forecasting gap of 7 days is enforced between the last available sensor observation and the target assessment. The 18 patients are randomly partitioned at the patient level into training (12 patients), validation (3 patients), and held-out test (3 patients) sets, ensuring no patient appears in more than one split.

4.2 Implementation Details

After performing the procedure described in the previous Sect. 3.2, we perform a grid search over the numbers of top-ranked sensor and demographic features, M and D, selected by SHAP ranking. The best-performing configuration is selected based on validation performance. The GRU encoder and prediction heads use a hidden dimension of 24. Dropout with probability 0.3 is applied in the input projection and prediction heads. Model parameters are optimized using AdamW with a learning rate of 2×10^{-3}. Training is performed for a maximum of 500 epochs with early stopping patience of 45 epochs and gradient clipping at a maximum norm of 1.0. All experiments use a fixed random seed. Additionally, the joint SIS and OHS prediction model is trained using four additional random seeds to assess performance stability and robustness to random initialization.

5 Results and Discussion

5.1 Overall Forecasting Performance

All reported results are generated using the leakage-safe training-only SHAP computation as stated in Sect. 3.2. Results from the best-performing seed are reported. Per-item and total-score performance (MAE, RMSE, R^2, Pearson r) on held-out test set (3 patients, 12 examples) are reported in Tables 1 and 2. Per-item metrics are pooled across all individual item predictions (12 items for OHS, 6 items for SIS), while total-score metrics are computed on reconstructed scores (sum of predicted items vs. sum of ground-truth items).

Table 1. Test-set performance of the multi-task GRU model trained jointly on OHS and SIS.

Outcome	Method	MAE	RMSE	R^2	Pearson r
OHS	Per-item	0.741	0.931	0.181	0.468
	Total	3.134	4.292	0.176	0.672
SIS	Per-item	1.164	1.410	−0.405	0.583
	Total	6.088	6.648	−1.985	0.829

Table 2. Test-set performance of the single-task GRU baselines: one model trained exclusively on OHS items and one trained exclusively on SIS items, each using only its task-specific SHAP-ranked features.

Outcome	Method	MAE	RMSE	R^2	Pearson r
OHS	Per-item	0.943	1.149	−0.248	0.374
	Total	7.987	8.840	−2.496	−0.434
SIS	Per-item	0.976	1.244	−0.093	0.399
	Total	4.026	5.275	−0.879	−0.141

The multi-task model achieves a per-item MAE of 0.741 for OHS (0–4 Likert scale) and 1.164 for SIS (1–5 Likert scale). OHS prediction attains positive R^2 (+0.181) and Pearson r (0.468), indicating meaningful predictive capability at the item level despite evaluation on only three unseen patients. In contrast, SIS is more difficult to predict, with negative R^2 (−0.405) but positive correlation ($r = 0.583$), suggesting the model captures relative trends while exhibiting substantial prediction error. At the total-score level, correlations improve for both outcomes ($r = 0.672$ for OHS and $r = 0.829$ for SIS), indicating that aggregation across questionnaire items reduces the impact of item-level noise and better reveals overall recovery trajectories.

To assess whether joint multi-task training confers an advantage over independent single-task models, we compare the multi-task GRU against two single-task baselines trained separately on OHS and SIS using only task-specific SHAP-ranked features (Table 2). For OHS, multi-task training substantially improves performance, reducing per-item MAE from 0.943 to 0.741 and increasing R^2 from -0.248 to $+0.181$. Total-score performance improves even more markedly, with MAE decreasing from 7.987 to 3.134 and Pearson r increasing from -0.434 to $+0.672$. For SIS, joint training yields a slight increase in per-item MAE (1.164 vs. 0.976), but total-score correlation improves dramatically from -0.141 to $+0.829$. These results suggest that shared supervision enables the encoder to learn recovery patterns that generalize better across patients, particularly at the clinically relevant total-score level.

Overall, the multi-task architecture produces substantially better generalization than the single-task baselines across nearly all metrics, with the most pronounced gains at the total-score level. Pearson correlations for both OHS and SIS is particularly compelling, suggesting that cross-task supervision is essential for reliable recovery forecasting in this small-cohort setting.

5.2 Per-Item Forecasting Performance

At the item level (per question prediction), performance shown in Table 3 is uneven across both outcomes, although OHS generally demonstrates stronger predictive performance than SIS. For OHS, ohs-10 achieves the strongest correlation with the ground truth ($r = 0.865$), followed by ohs-03 ($r = 0.851$) and ohs-02 ($r = 0.701$), indicating that the model captures the relative variation of these recovery dimensions reasonably well. In contrast, ohs-12 achieves the highest R^2 (0.414), suggesting the most accurate reconstruction of absolute values. Several items, including ohs-02, ohs-08, ohs-11, and ohs-12, obtain positive R^2 values, demonstrating meaningful predictive signal beyond a mean predictor. However, performance is not uniformly strong. Items ohs-05 and ohs-06 show negative correlations ($r = -0.677$ and $r = -0.266$, respectively) and negative R^2 values, suggesting that the model fails to capture the underlying direction of change for these aspects of hip function. Notably, ohs-05 exhibits one of the lowest MAEs (0.394) despite its negative correlation, indicating that predictions remain numerically close to the target values while failing to preserve patient-specific variation.

For SIS, item-level forecasting is generally weaker and more inconsistent. All six SIS items produce negative R^2 values, indicating that none outperform a mean-value baseline in terms of explained variance. Nevertheless, several items maintain moderate-to-strong positive correlations, including sis-05 ($r = 0.771$), sis-04 ($r = 0.757$), and sis-02 ($r = 0.409$), suggesting that the model captures relative ranking or temporal trends despite substantial prediction error. In contrast, sis-03 exhibits a negative correlation ($r = -0.422$), indicating an inverse relationship between predictions and ground truth. The poorest overall performance is observed for sis-06, which records the largest MAE (1.871), the largest RMSE (2.120), and a strongly negative R^2 (-1.568), suggesting that this recovery dimension is not well represented by the available sensor features. Overall, the SIS results reveal a disconnect between correlation and explained variance. Although aggregation substantially increases Pearson correlation ($r = 0.829$), the strongly negative total-score R^2 (−1.985) suggests that these trends remain poorly calibrated in magnitude. Consequently, aggregation appears to amplify systematic biases that accumulate when item predictions are summed rather than reduce prediction error.

Table 3. Per-item test-set forecasting performance for OHS (0–4 scale) and SIS (1–5 scale). Metrics are computed across all test examples ($n = 12$).

Outcome	Item	MAE	RMSE	R^2	Pearson r
OHS	ohs-01	1.035	1.130	−0.045	0.005
OHS	ohs-02	1.035	1.142	0.243	0.701
OHS	ohs-03	1.034	1.252	0.059	0.851
OHS	ohs-04	0.772	0.925	−0.061	0.471
OHS	ohs-05	0.394	0.565	−0.316	−0.677
OHS	ohs-06	0.563	0.632	−0.799	−0.266
OHS	ohs-07	0.931	1.098	−0.015	0.127
OHS	ohs-08	0.409	0.571	0.374	0.635
OHS	ohs-09	0.452	0.570	0.083	0.567
OHS	ohs-10	0.830	1.084	0.055	0.865
OHS	ohs-11	0.904	1.114	0.289	0.586
OHS	ohs-12	0.527	0.634	0.414	0.653
SIS	sis-01	0.955	1.064	−1.175	0.247
SIS	sis-02	0.734	0.892	−0.688	0.409
SIS	sis-03	0.979	1.130	−0.999	−0.422
SIS	sis-04	1.115	1.211	−1.819	0.757
SIS	sis-05	1.325	1.662	−0.118	0.771
SIS	sis-06	1.871	2.120	−1.568	0.101

5.3 Prediction Distribution Analysis

To further investigate model behaviour, predicted scores are compared directly against ground-truth clinical assessment scores. The predicted scores from train set and test set are shown in Fig. 3 and Fig. 4.

The training predictions in Fig. 3 follow the diagonal behaviour reasonably well for both OHS and SIS, confirming that the model has learned a meaningful signal from the training patients. SIS predictions appear particularly well aligned with the diagonal across the full score range, while OHS exhibits greater dispersion and a tendency to underpredict at higher ground-truth values. This pattern is most evident for patients with OHS scores above 35, where predictions remain compressed toward the centre of the distribution. Such behaviour is expected in small datasets, where extreme examples are underrepresented and the SmoothL1 objective encourages conservative predictions closer to the population mean.

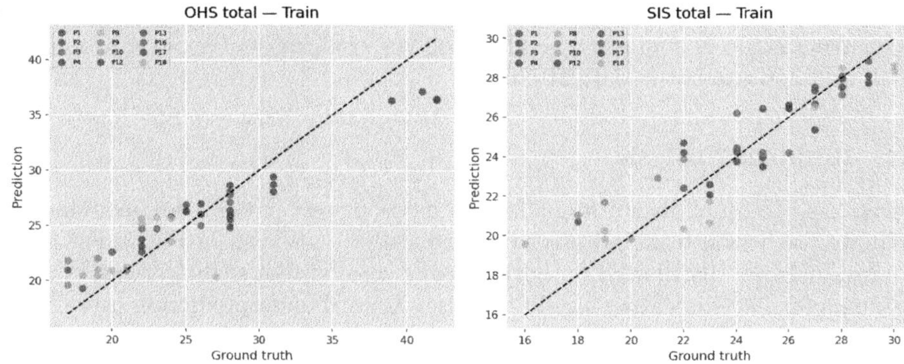

Fig. 3. Predicted versus ground-truth values for OHS and SIS from train set. The diagonal line represents ideal prediction performance.

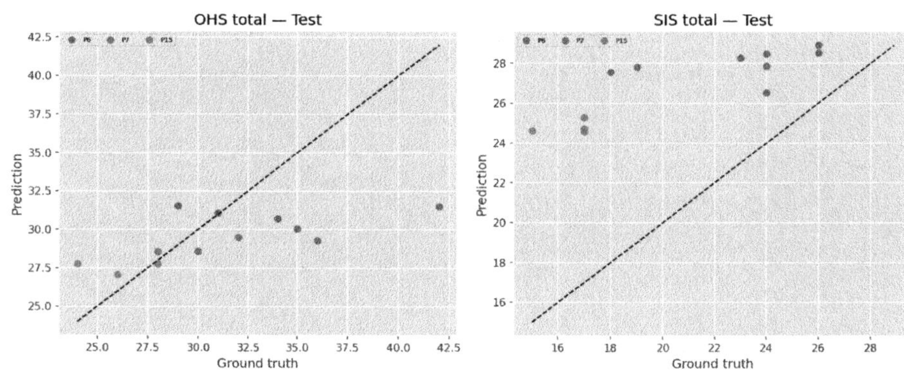

Fig. 4. Predicted versus ground-truth values for OHS and SIS from test set. The diagonal line represents ideal prediction performance.

For OHS, the test predictions in Fig. 4 exhibit a clear compression effect, with predictions concentrated between approximately 27 and 32 despite a much wider ground-truth range. Lower scores tend to be overpredicted and higher scores consistently underpredicted, producing a positive Pearson correlation ($r = 0.672$) but only modest explanatory power ($R^2 = 0.176$). The effect is even more pronounced for SIS, where predictions cluster near 25–29 regardless of the ground-truth value, resulting in systematic overprediction at lower scores and weaker calibration. Nevertheless, the model preserves relative patient ordering, achieving a strong total-score correlation ($r = 0.829$) despite a highly negative R^2 (-1.985). For OHS, P15 (purple) exhibits the closest alignment to the diagonal, whereas for SIS, all three show larger systematic offsets. This indicates that the recovery trajectory for OHS is more consistent with patterns represented in the training cohort, while predictions for SIS exhibit greater patient-specific variability that is more difficult for the model to generalize. Together, these results suggest that the model captures broad recovery trends but struggles to accurately reconstruct score magnitudes, highlighting the challenges of generalization in a small and heterogeneous patient cohort.

5.4 Residual Analysis

Normalized residual distributions for the held-out test patients are examined to evaluate prediction bias and error characteristics, as shown in Fig. 5. For OHS, the residuals are distributed on both sides of zero, with a slightly positive mean residual of 0.44σ, indicating a mild tendency toward underprediction overall. For OHS, P7 (blue) residuals are showing underprediction, while P6 (red) and P15 (purple) residuals are showing overprediction. These opposing biases partially cancel, producing the near-zero mean. Although the overall bias is small, the spread of residuals suggests that prediction accuracy remains strongly dependent on the individual patient.

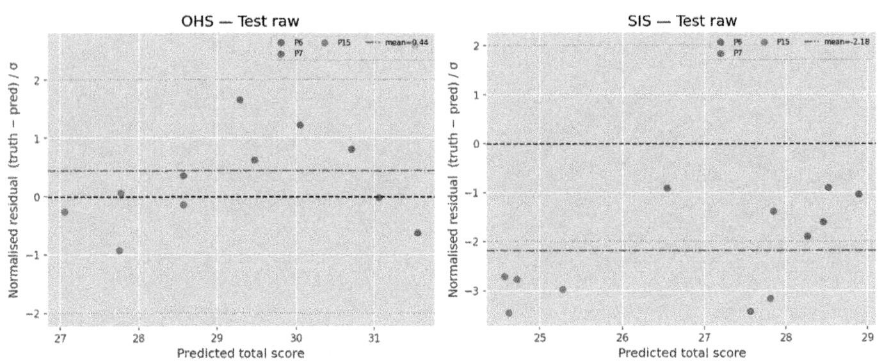

Fig. 5. Residual distributions for OHS and SIS predictions. Residuals are computed as ground truth minus prediction and normalized by σ. Dashed lines indicate mean residual values.

For SIS, the residual distribution exhibits a pronounced negative bias, with a mean residual of -2.18σ that indicates systematic overprediction of SIS total scores. This effect is particularly evident for P15 and P7, whose residuals frequently fall below -2σ, while P6 remains the least biased. Unlike OHS, there is little evidence of residual cancellation across patients, as the error direction is largely consistent throughout the test set. Together, these results indicate that the model overestimates SIS recovery while showing mild underprediction for OHS. The patient-specific clustering of residuals suggests that generalization errors arise primarily from differences in individual recovery trajectories rather than from random prediction noise, reinforcing the need for larger and more diverse training populations.

5.5 Performance Stability Across Five Random Seeds

To assess robustness, the full pipeline was repeated across five random seeds with the same held-out test set. For each seed, SHAP feature ranking, feature-selection grid search, and model training were performed independently. The resulting test predictions were also averaged to form a seed-level ensemble, with performance summarized in Table 4. Across individual seeds, performance exhibits noticeable variability for both OHS and SIS, which reflects sensitivity to random initialization and feature-selection decisions within the pipeline. For OHS, MAE ranges from 0.741 to 0.943 and Pearson correlation ranges from −0.014 to 0.468, indicating that model performance is sensitive to the feature rankings and parameter estimates produced by different random initializations. Notably, S1 achieves substantially lower MAE and higher R^2 than all other seeds. Similarly, SIS performance varies across seeds, with Pearson correlations ranging from 0.169 to 0.583 and R^2 values fluctuating between −0.405 and 0.050. Despite this variability, error metrics remain relatively stable across seeds, suggesting that the model produces broadly similar score estimates while differing in its ability to capture patient-specific variation.

The ensemble results demonstrate a modest benefit of aggregating predictions across multiple independently trained models. For both OHS and SIS, the ensemble reduces prediction error relative to the average individual seed. Although the ensemble does not fully resolve the challenges associated with the small and heterogeneous patient cohort, averaging across seeds mitigates some of the variability introduced by random initialization and feature-selection decisions. These findings suggest that the model captures reproducible recovery-related signals, but that performance remains constrained by the limited size and diversity of the available dataset.

Table 4. Per-item test-set performance across five random seeds and the ensemble for the multi-task GRU model. S = seed; Ens. = ensemble.

Outcome	Metric	S1	S2	S3	S4	S5	Ens.
OHS	MAE	0.741	0.921	0.918	0.943	0.923	0.860
	RMSE	0.931	1.110	1.093	1.113	1.098	1.014
	R^2	0.181	−0.165	−0.129	−0.170	−0.138	0.028
	Pearson r	0.468	0.296	−0.014	0.075	0.269	0.361
SIS	MAE	1.164	1.034	1.036	0.903	0.941	0.960
	RMSE	1.410	1.318	1.320	1.159	1.191	1.216
	R^2	−0.405	−0.228	−0.231	0.050	−0.002	−0.044
	Pearson r	0.583	0.330	0.461	0.447	0.169	0.573

6 Limitations

A primary limitation is the small dataset size. With only 18 patients and a 12/3/3 train/validation/test split, the GRU is susceptible to overfitting patient-specific temporal patterns rather than learning robust population-level representations, as evidenced by the gap between training and test performance. The 3-patient test set further allows a single atypical individual to disproportionately influence aggregate metrics, limiting confidence in generalization. The GRU encoder also processes sequences unidirectionally and compresses long-term dependencies into a low-dimensional hidden state, potentially limiting its ability to capture complex interactions between sensor modalities over extended rehabilitation periods. Although the SIS and OHS prediction heads are task-specific, they share the same encoder, which may cause negative transfer when the temporal dynamics of the two outcomes differ substantially. Finally, training relies solely on SmoothL1 loss without explicit regularization for temporal consistency or uncertainty calibration. Although dropout and weight decay are applied, these mechanisms alone may be insufficient for stabilizing learning under small-sample conditions.

7 Conclusion

This paper presented a leakage-safe multi-task temporal forecasting framework for predicting future OHS and SIS outcomes from longitudinal multimodal sensor data collected during LLF recovery. By combining SHAP-guided feature selection, learnable feature gating, a shared GRU encoder, and task-specific recency-aware attention, the model captured broad recovery-related patterns from variable-length patient histories. Despite the small cohort size and substantial patient heterogeneity, joint prediction of OHS and SIS improved generalization over independently trained single-task models, particularly for clinically relevant total-score forecasting and correlation-based measures. While the model successfully captured overall recovery trends, prediction calibration and robustness remained limited, highlighting the challenges of longitudinal forecasting in small rehabilitation datasets. These findings demonstrate the feasibility of forecasting functional and psychosocial recovery from passive sensing data and provide a foundation for future personalized rehabilitation monitoring systems.

8 Future Work

The primary future direction is improving generalization under limited patient data. The current results suggest that the model partially overfits. Future work should therefore prioritize diverse longitudinal datasets, particularly through multi-site data collection and external validation cohorts. Such evaluation is necessary to determine whether the learned temporal representations remain robust across different rehabilitation populations, sensor acquisition conditions, and clinical workflows. From a modelling perspective, future research should focus on replacing the current GRU-based architecture with more expressive temporal learning frameworks. Transformer-based or multi-scale temporal encoders may better capture long-range rehabilitation dynamics and heterogeneous recovery trajectories than the current sequential hidden-state formulation. In parallel, self-supervised pretraining on raw wearable sensor streams could reduce dependence on handcrafted statistical features and improve representation robustness in low-data clinical settings.

Acknowledgements. This work was supported by Mitacs Globalink Research Award and the Innovative Cognitive Computing Research Center.

References

1. Abedi, A., Chu, C.H., Khan, S.S.: Multimodal sensor dataset for monitoring older adults post lower limb fractures in community settings. Sci. Data **12**(1), 733 (2025)
2. Choi, E., Bahadori, M.T., Sun, J., Kulas, J., Schuetz, A., Stewart, W.: Retain: an interpretable predictive model for healthcare using reverse time attention mechanism. Adv. Neural. Inf. Process. Syst. **29** (2016)
3. Chung, J., Gulcehre, C., Cho, K., Bengio, Y.: Empirical evaluation of gated recurrent neural networks on sequence modeling. arXiv preprint arXiv:1412.3555 (2014)
4. Dakhil, S., et al.: Longitudinal trajectories of functional recovery after hip fracture. PLoS ONE **18**(3), e0283551 (2023)
5. Davis, S.E., Matheny, M.E., Balu, S., Sendak, M.P.: A framework for understanding label leakage in machine learning for health care. J. Am. Med. Inf. Assoc. **31**(1), 274–280 (2024)
6. Dawson, J., Fitzpatrick, R., Carr, A., Murray, D.: Questionnaire on the perceptions of patients about total hip replacement. J. Bone Joint Surg. Brit. **78**(2), 185–190 (1996)
7. Dayyani, F., Chu, C.H., Abedi, A., Khan, S.S.: Correlations between social isolation and functional decline in older adults after lower limb fractures using multimodal sensors: A pilot study. Algorithms **17**(9), 383 (2024)
8. Graves, A.: Long short-term memory. In: Supervised Sequence Labelling with Recurrent Neural Networks, pp. 37–45 (2012)
9. He, Y., Zhao, J.: Temporal convolutional networks for anomaly detection in time series. In: Journal of Physics: Conference Series, vol. 1213, p. 042050. IOP Publishing (2019)
10. Kapoor, S., Narayanan, A.: Leakage and the reproducibility crisis in machine-learning-based science. Patterns **4**(9) (2023)
11. Khan, S.S., Gu, T., Spinelli, L., Wang, R.H.: Sensor-based assessment of social isolation in community-dwelling older adults: a scoping review. Biomed. Eng. Online **22**(1), 18 (2023)
12. King, S.W., et al.: Wearable sensors to guide remote rehabilitation following knee arthroplasty surgery. Indian J. Orthopaedics **57**(5), 624–634 (2023)
13. Lundberg, S.M., Lee, S.I.: A unified approach to interpreting model predictions. Adv. Neural Inf. Process. Syst. **30** (2017)
14. Murray, D., et al.: The use of the oxford hip and knee scores. J. Bone Joint Surg. Brit. **89**(8), 1010–1014 (2007)

15. Nicholson, N.R., Jr., Feinn, R., Casey, E., Dixon, J.: Psychometric evaluation of the social isolation scale in older adults. Gerontologist **60**(7), e491–e501 (2020)
16. Pliannuom, S., et al.: Characteristics and effects of home-based digital health interventions on functional outcomes in older patients with hip fractures after surgery: systematic review and meta-analysis. J. Med. Internet Res. **26**, e49482 (2024)
17. Porciuncula, F., et al.: Wearable movement sensors for rehabilitation: a focused review of technological and clinical advances. PM&R **10**(9), S220–S232 (2018)
18. Vaswani, A.,.: Attention is all you need. Adv. Neural Inf. Process. Syst. **30** (2017)
19. Veronese, N., Kolk, H., Maggi, S.: Orthogeriatrics: the management of older patients with fragility fractures (2021)

AI for Aging and Intelligent Assisted Living

Small Language Models for Early Alzheimer's Risk Detection: Accuracy-Efficiency Trade-Offs in the DMV Framework

Vinatha Viswanathan, Peng Zhang, and Divya Chaudhary(✉)

Khoury College of Computer Sciences, Northeastern University, Seattle, USA
{viswanathan.vin,zhang.peng2,d.chaudhary}@northeastern.edu

Abstract. Early risk detection of Alzheimer's Disease (AD) is essential for maximizing the efficacy of clinical interventions. While Large Language Models (LLMs) achieve high accuracy within the Data Processing-Model Training-Validation (DMV) Framework, they present significant barriers to clinical deployment due to immense computational and memory overhead. This paper extends the DMV Framework by investigating the efficacy of Small Language Models (SLMs) as embedding engines, conducting a comparative analysis between encoder-only and decoder-only architectures for AD diagnosis. We further explore embedding quality for various quantization levels of these SLMs. Our findings reveal that both encoder-only and decoder-only models achieve a predictive accuracy improvement of 3.07% over the cuML Random Forest Regressor baseline. Our analysis also shows that encoder-only models require up to 72% less GPU memory compared to the decoder-only models examined in this study. Overall, we demonstrate that 4-bit quantization preserves over 99.98% of 16-bit embedding baselines while reducing GPU memory footprint by an average of 16%.

Keywords: Alzheimer's Disease · Small Language Models · Quantization · Embedding · Random Forest · DMV Framework

1 Introduction

Alzheimer's Disease (AD) is a progressive neurodegenerative disorder marked by a long preclinical phase where linguistic shifts in reduced vocabulary and loss of semantic coherence serve as vital "digital biomarkers" of early cognitive decline [1]. LLMs offer an unprecedented capability to capture these complex contextual dependencies. Advances in Natural Language Processing (NLP), particularly the recent development of LLMs, have enabled the accurate modeling of high-dimensional semantic relationships within speech and text [2]. However, integrating such models into routine clinical screening pipelines remains challenging due to their substantial operational demands, including large memory

S. S. Khan et al. (Eds.): IJCAI 2026, CCIS 3127, pp. 47–62, 2027.
https://doi.org/10.1007/978-981-92-5190-2_4

footprints, high computational power requirements, and network latency overhead [3]. Consequently, SLMs typically defined as models with fewer than 10 billion parameters [4], have emerged as a practical alternative. SLMs offer significantly lower computational costs and improved efficiency while preserving sufficient representational capacity for meaningful linguistic feature extraction [5].

In this study, we propose a hybrid architecture that leverages SLMs to generate embeddings that complement traditional structured features. By utilizing both encoder-only and decoder-only architectures, we extract high-dimensional semantic vectors that are then concatenated with standard features to train a Random Forest Regressor. This approach seeks to combine the contextual intelligence of transformer-based architectures with the non-linear partitioning of a Random Forest Regressor. This hybrid method allows for the seamless inclusion of linguistic features without the prohibitive GPU memory overhead and hardware costs associated with high-parameter LLMs.

A critical component of this study is the investigation of the trade-off between model precision and operational efficiency. Drawing inspiration from recent findings in bit-level scaling laws [6], we evaluate our SLM-based feature extractors across varying quantization levels, specifically 4-bit, 8-bit, and 16-bit precisions. Our objective is to analyze how reducing the bit-width of model parameters impact the quality of resulting embeddings and subsequently the final regression performance. By quantifying the relationship between memory footprint and predictive accuracy, we seek to identify an optimal "sweet spot" where model compression allows for maximum efficiency without compromising the overall integrity of the feature space.

2 Related Work

2.1 Computational Methods for AD Detection

Early research on AD detection focused on identifying impairments in patient speech and transcripts. Initially, the approaches relied on manually extracted linguistic features and traditional statistical classifiers such as logistic regression [7] and support vector machines [8] to distinguish AD patients from healthy controls [9]. Subsequent studies adopted machine learning methods, including ensemble approaches such as random forests and gradient boosting, to improve the detection of subtle linguistic and acoustic markers [10]. With the availability of larger datasets, deep learning techniques such as Convolutional Neural Networks (CNNs) [11] and Recurrent Neural Networks (RNNs) [12] were used to capture abnormal lexical patterns and cognitive impairments.

More recently, transformer-based pretrained language models have been applied to speech transcripts and clinical texts [13]. Pretrained on large corpora, these models capture contextual relationships and discourse-level features such as semantic coherence, lexical diversity, syntactic structure, and contextual word usage, that are often impaired in AD patients. Building on this progress, recent studies integrate LLMs into machine learning pipelines as feature generators, combining clinical text representations with auxiliary signals (e.g., geolocation data) to improve AD detection by capturing both cognitive, linguistic and behavioral indicators [14]. However, the large size of LLMs pose practical challenges in clinical settings as evaluation extends beyond raw predictive performance to include robustness and reliability, while also requiring cost-aware and efficient assessment methods applicable across languages and clinical scenarios [15]. This study therefore explores the applicability and efficacy of SLM embeddings within the model training pipeline for AD risk detection.

2.2 Quantization Paradigms

The landscape of post-training quantization for SLMs is divided into two primary categories, one-shot methods and zero-shot methods [16]. One-shot techniques, represented by GPTQ [17] and Activation-Aware Weight Quantization (AWQ) [18], utilize a calibration mini-batch to minimize rounding errors through the analysis of second-order information or activation distributions. Although these methods achieve high fidelity, they introduce a significant optimization tax often requiring multiple hours of computation [6]. Conversely, zero-shot methods such as Round-To-Nearest (RTN) [19] facilitate immediate model initialization by bypassing the data-intensive calibration phase entirely. In this study, we compress weight tensors to lower bit-widths while enabling SLMs to efficiently operate within the stringent memory constraints of edge devices and consumer-grade hardware. We systematically study the quantization-induced degradation of SLM embeddings and analyze the reduction in GPU utilization at lower precisions.

3 SLM Integrated DMV Framework

The SLM Integrated DMV framework, illustrated in Fig. 1, effectively enhances the standard Data Processing-Model Training-Validation pipeline by leveraging these compact models as efficient feature extractors. Unlike LLMs, SLMs achieve a superior balance between computational overhead and feature richness owing to their low-parameter footprint. Integrating these models directly into our DMV framework maintains high-dimensional data representation while keeping the pipeline accessible for resource-constrained clinical environments.

The pipeline begins with a data processing stage, where raw clinical data is preprocessed, and bifurcated into categorical and numerical features for encoding and normalization. We then implement a SLM integration layer that utilizes both Encoder-only and Decoder-only architectures to extract high-dimensional embeddings across three distinct quantization levels (4-bit, 8-bit, 16-bit). These embeddings are concatenated with the static features to serve as inputs to a cuML Random Forest Regressor[1]. Finally, the framework is evaluated through a three stage validation suite that assesses Predictive Accuracy, Quantization Degradation, and Memory Footprint to quantify the trade-offs between model compression and diagnostic reliability.

---

[1] https://github.com/rapidsai/cuml.

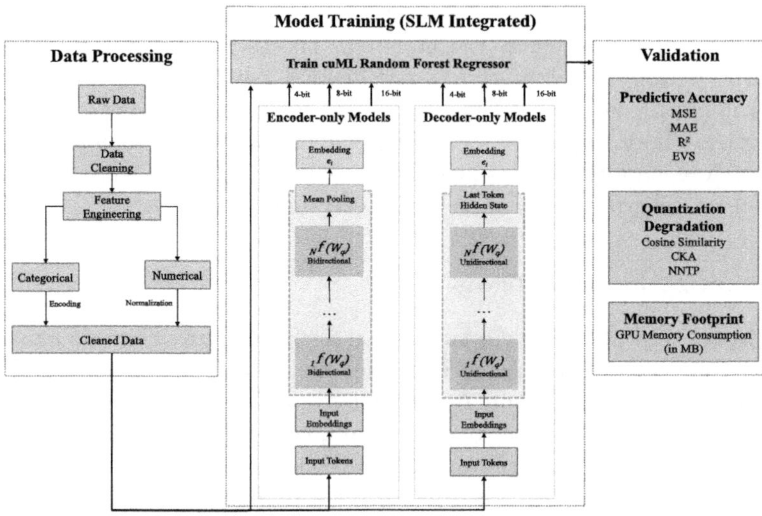

Fig. 1. Overview of the SLM Integrated DMV Framework for AD Risk Detection.

4 Evaluation

We empirically validate the effectiveness of our extended DMV framework by carefully examining its performance across SLMs of different architectural types (e.g. encoder-only, decoder-only), model sizes (14M to 3.8B parameters) and varying quantization levels (4-bit, 8-bit, 16-bit).

4.1 Experimental Setup

Dataset. We evaluate SLM embeddings on the Centers for Disease Control and Prevention (CDC) dataset[2], which comprises 284K records across 31 health and demographic fields. This dataset tracks behavioral, environmental, and clinical risk factors at the population level that are epidemiologically linked to AD onset and progression. Rather than containing individual patient diagnoses, it captures area-level health indicators such as rates of cognitive decline, physical inactivity, obesity, and mental distress that public health authorities use to characterize AD risk across communities.

In order to capture semantic structure, we construct a combined text feature by concatenating question, class, topic, and two stratification fields, embedded using encoder and decoder-only SLMs. These pretrained language model embeddings capture semantic relationships across the dataset's 39 health topics and demographic stratifications [20], providing the downstream Random Forest Regressor with a representation that generalizes across categories rather than treating them as independent dimensions. The prediction target is the

[2] https://data.cdc.gov/resource/hfr9-rurv.

`data_value` field, a continuous numerical variable representing population-level health statistics such as the percentage of older adults reporting frequent mental distress or currently classified as obese, aggregated across geographic and demographic strata. We frame this as a regression task because the target is an unbounded continuous rate or percentage, not a discrete diagnostic category. Accurate prediction of these scores enables identification of high-risk population segments, supporting early intervention at the community level. The scores represent the risk or possibility of developing Alzheimer's disease for the patients in this area. The higher the score, the more likely the patients in this area are to have Alzheimer's disease. Following [30], we employ 10-fold cross-validation, partitioning the dataset into 10 equal stratified folds. In each iteration, nine folds are used for training and one for evaluation, with all reported metrics averaged across folds. This ensures a robust and reproducible estimate of generalisation performance.

Models. We conduct our experiments across the encoder-only (DistilBERT-base [21], DeBERTa-v3-base, ELECTRA-small, XLM-RoBERTa-base [22], MiniLM-L6-v2, MPNet-base-v2 [23], MiniLM-L12-v2, BioBERT-v1.1) and decoder-only (LLaMA-3.2-3B [24], Phi-3.5-mini [25], Qwen2.5-3B [26,27], Gemma-2B, Gemma-2B-IT, SmolLM3-3B [28], SmolVLM-256M-Instruct, SmolLM2-135M [29] models.

Baselines. We adopt the GPU-accelerated cuML Random Forest Regressor as our baseline, replacing the scikit-learn implementation from prior work [30], which utilized LLMs as feature extractors coupled with a Scikit-Learn Random Forest Regressor for AD diagnosis.

Quantization Configuration. Each SLM is evaluated at three precision levels: 4-bit, 8-bit, and 16-bit, with 16-bit serving as the unquantized reference for representational fidelity analysis. All quantization is performed using Round-To-Nearest (RTN), a zero-shot post-training method that requires no calibration data and is applied identically across all models and precision levels. RTN is selected for its simplicity and reproducibility, enabling a clean comparison of quantization-induced degradation without introducing confounds from calibration set composition. Weight tensors are compressed to the target bit-width at inference time, with the 16-bit baseline retained for embedding similarity comparisons.

Evaluation Metrics. We evaluate across three dimensions. Predictive accuracy is measured using Mean Squared Error (MSE), Mean Absolute Error (MAE), R-squared (R^2), and Explained Variance Score (EVS) [31]. Quantization degradation is assessed via Cosine Similarity, Centred Kernel Alignment (CKA), and Nearest Neighbor Topology Preservation (NNTP) to capture geometric distortion in embedding space. Memory footprint is reported as GPU memory consumption (in MB) per embedding batch.

4.2 Baseline Training Using CuML

Across all hyperparameter sweeps, cuML demonstrates clear and largely monotonic improvements as model capacity increases, particularly with respect to `max_depth` and `n_estimators`. Increasing `max_depth` from 22 to 34 yields the most substantial gain, reducing MSE by 35% and improving R^2 by 1.6%. Similarly, performance improves steadily as `n_estimators` increases from 600 to 1200, after which gains plateau at 1300 and 1500, indicating diminishing returns beyond 1200 trees. For `max_features`, a moderate value of 0.65 outperforms both lower and higher settings, suggesting that controlled feature subsampling provides beneficial regularization while preserving predictive signal [32]. Adjusting `min_samples_leaf` from 1 to 2 yields a consistent improvement, with no further benefit observed for larger values [33]. As summarized in Table 1, the best-performing cuML configuration achieves predictive performance close to the Scikit-learn baseline, while delivering a 1.5 times reduction in training time.

Table 1. Hyperparameter and performance metrics for Scikit-learn and cuML.

Hyperparameters	Scikit-learn	cuML
`n_estimators`	100	1200
`max_depth`	None	32
`max_features`	1.0	0.65
`min_samples_leaf`	1	2
Metric	**Scikit-learn**	**cuML**
Time	3523.12 s	2437.53 s
MSE	2.3549e-2	3.0115e-2
MAE	8.5575e-2	1.0787e-1
R^2	0.9766	0.9700
EVS	0.9766	0.9700

4.3 Analysis of Hyperparameter Tuning

A systematic hyperparameter study was conducted to identify the configuration that maximized predictive performance while maintaining computational efficiency, consistent with established findings on variance reduction and bias-variance trade-offs in ensemble learning [32].

Effect of `n_estimators`. Figure 2(a) shows how increasing ensemble size steadily improves accuracy through variance reduction, stabilizing around 1200 trees, beyond which gains are marginal while training time increases linearly.

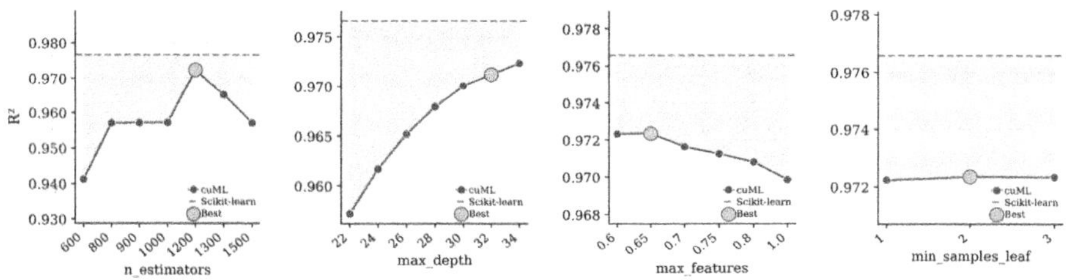

(a) Effect of Tree Count (b) Effect of Tree Depth (c) Effect of Feature Ratio (d) Effect of Leaf Sample Size

Fig. 2. Effect of hyperparameter tuning on predictive accuracy across model parameters. Comparative analysis of R^2 performance between cuML and Scikit-learn Random Forest Regressor.

Effect of `max_depth`. Figure 2(b) shows the effect of tree depth. Increasing depth improves performance by enabling the model to capture higher-order interactions among predictors. Accuracy saturates near depth 32, while training time increases sharply beyond this point.

Effect of `max_features`. Figure 2(c) presents the effect of feature subsampling. Intermediate subsampling (`max_features` = 0.65) achieves the highest performance. Using all features at each split slightly degrades accuracy, likely due to increased correlation among trees and reduced ensemble diversity.

Effect of `min_samples_leaf`. Figure 2(d) shows the effect of leaf size. Setting `min_samples_leaf` to 2 improves generalization compared with fully grown leaves by stabilizing terminal node predictions. Larger values oversmooth the model and reduce accuracy.

Effect of `n_bins` and `min_samples_split`. Increasing `n_bins` beyond 128 did not improve accuracy but increases training time. Variations in `min_samples_split` had negligible impact under the chosen leaf constraint, as most nodes already contained sufficient samples.

4.4 Experiment on SLM Embeddings

The experimental results for the SLM-embedded Random Forest Regressor across various quantization levels (4-bit, 8-bit, and 16-bit) are presented in Table 2. Each model generates a fixed-length sentence embedding by applying mean pooling over the final hidden states, using the attention mask to ignore padding tokens. For encoder-only models, this captures the sentence representation directly, while for decoder-only models, it yields a fixed-length vector from sequential token states [20]. The resulting embedding vectors are concatenated with the tabular feature matrix to form a unified feature set, which is used to train a cuML Random Forest Regressor. Integrating contextual embeddings with structured features has shown to enhance predictive performance in tabular

Table 2. Performance evaluation of SLM-embedded Random Forest Regressor at three quantization levels (4-bit, 8-bit, and 16-bit), compared across eight encoder-only and eight decoder-only models.

Model	4-bit					8-bit					16-bit				
	MSE ($\times10^{-4}$)	MAE ($\times10^{-3}$)	R^2	EVS	GPU (MB)	MSE ($\times10^{-4}$)	MAE ($\times10^{-3}$)	R^2	EVS	GPU (MB)	MSE ($\times10^{-4}$)	MAE ($\times10^{-3}$)	R^2	EVS	GPU (MB)
Encoder-Only Models															
DistilBERT-base	1.754296	5.886351	0.999825	0.999825	2300.56	1.795502	6.114147	0.999821	0.999821	2512.56	1.731668	5.882802	0.999827	0.999827	1216.56
DeBERTa-v3-base	1.911973	6.020593	0.999809	0.999809	1458.56	1.963464	6.190703	0.999804	0.999804	1792.56	1.949271	6.022549	0.999806	0.999806	3256.72
ELECTRA-small	1.788596	5.798482	0.999822	0.999822	2362.48	1.879415	5.990378	0.999813	0.999813	3260.56	1.818634	5.809357	0.999819	0.999819	2242.56
XLM-RoBERTa-base	1.774798	5.954140	0.999823	0.999823	3756.56	1.849404	6.155289	0.999816	0.999816	3748.56	1.787775	5.905358	0.999822	0.999822	2704.55
MiniLM-L6-v2	1.641918	5.844989	0.999836	0.999836	2421.40	1.641447	6.034671	0.999836	0.999836	3276.39	1.576917	5.890145	0.999820	0.999843	2310.55
MPNet-base-v2	1.758830	5.988932	0.999825	0.999825	2594.56	1.764177	6.145327	0.999824	0.999824	3392.56	1.761072	5.973564	0.999824	0.999824	2410.54
MiniLM-L12-v2	1.633220	5.839337	0.999837	0.999837	2446.56	1.606001	6.038825	0.999840	0.999840	3322.56	1.556723	5.882549	0.999845	0.999845	1144.39
BioBERT-v1.1	1.805142	5.898984	0.999820	0.999820	3390.56	1.789894	6.124232	0.999821	0.999821	2620.56	1.742710	5.894899	0.999826	0.999826	2412.56
Decoder-Only Models															
LLaMA-3.2-3B	1.622366	6.065927	0.999838	0.999838	7056.55	1.607994	6.221108	0.999840	0.999840	8324.55	1.634889	6.052704	0.999837	0.999836	8972.77
Phi-3.5-mini	1.600486	6.171901	0.999840	0.999840	8900.56	1.614317	6.218579	0.999839	0.999839	9224.56	1.546536	6.159768	0.999846	0.999846	9690.02
Qwen2.5-3B	1.897729	6.080143	0.999811	0.999811	6715.12	1.877550	6.182346	0.999813	0.999813	6833.20	1.857439	5.997868	0.999815	0.999815	7386.05
Gemma-2B	1.953170	6.138557	0.999805	0.999805	5746.33	1.956613	6.184450	0.999805	0.999805	6975.71	1.926347	6.127510	0.999808	0.999808	6619.85
Gemma-2B-IT	1.885146	6.118815	0.999812	0.999812	6814.32	2.007307	6.184369	0.999800	0.999800	7004.29	1.934716	6.123312	0.999807	0.999807	8518.54
SmolLM3-3B	1.896520	6.072302	0.999811	0.999811	8004.56	1.979574	6.213207	0.999803	0.999803	8932.55	1.833749	6.049251	0.999817	0.999817	7116.26
SmolVLM-256M-Instruct	1.540442	5.978021	0.999846	0.999846	3468.28	1.540642	6.069364	0.999846	0.999846	3679.25	1.459223	5.813720	0.999854	0.999854	2705.89
SmolLM2-135M	1.470911	5.973848	0.999853	0.999853	1361.41	1.547032	6.058625	0.999846	0.999846	1432.41	1.500413	5.815617	0.999850	0.999850	2476.15

learning settings [34]. We evaluate several SLMs across three quantization levels, 4-bit, 8-bit, and 16-bit.

All evaluated models substantially outperform the baseline, achieving average reductions of approximately 99.43% and 94.47% in MSE and MAE respectively, while improving R^2 and EVS scores by 3.07% over the baseline. Encoder-only models demonstrate a strong advantage in memory efficiency, with DistilBERT-base, DeBERTa-v3-base, and MiniLM-L12-v2 requiring approximately 72%, 70%, and 68% less GPU memory respectively compared to the median decoder-only model's memory footprint of 6,990 MB across all quantization levels. While decoder-only models maintain a consistent marginal lead in absolute precision across all quantization levels, the performance delta between the two architectural families remains statistically insignificant. To confirm this, we apply a paired Wilcoxon signed-rank test comparing mean R^2 scores across all evaluated models between the two architectural families, yielding $p = 0.875$. This convergence suggests a predictive parity for the embedding task, regardless of the underlying model. This is further underscored by the negligible degradation under 4-bit quantization. For each model, the retention ratio is computed as $(R^2_{\text{4-bit}}/R^2_{\text{16-bit}}) \times 100$, across all 16 evaluated models this ratio exceeds 99.98%, indicating that bit-depth reduction can be achieved without compromising on predictive accuracy.

4.5 Embedding Similarity Under Quantization

Vector-Level Similarity. To evaluate how well quantized models replicate 16-bit embeddings at the vector level, we compute the cosine similarity [35] between each 16-bit embedding and its corresponding quantized embedding for every record shown in Fig. 3(a) and (b). While 8-bit quantization maintains near-perfect similarity at half the memory footprint of 16-bit, 4-bit quantization offers a more aggressive reduction to one-quarter of the memory footprint, making it the primary focus of our analysis. On average, decoder-only models in this study maintain higher embedding similarity than encoder-only models at 4-bit, averaging at 97.1% and 93.3%, respectively.

Encoder-only models under 8-bit quantization maintain a 99.9% mean similarity compared to the 16-bit representations across all tested models. However, performance at 4-bit varies significantly. DeBERTa-v3-base exhibits the most substantial degradation, with mean similarity falling to 65.2%. In contrast, XLM-RoBERTa-base retains 99.8% mean similarity. Decoder-only models demonstrate broad resilience to 4-bit compression, with six out of eight tested models maintaining 97.2% mean similarity. In contrast, SmolVLM-256M-Instruct exhibits a mean similarity drop to 89.6%.

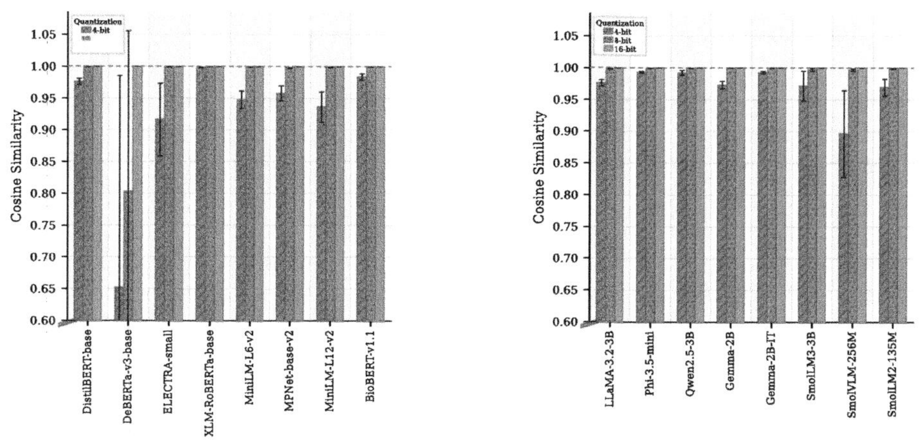

(a) Vector-Level Similarity for Encoder-Only Models

(b) Vector-Level Similarity for Decoder-Only Models

Fig. 3. Effect of quantization on vector-level representation fidelity. A comparative analysis of mean similarity between 16-bit benchmarks and quantized 4-bit and 8-bit variants across encoder-only and decoder-only models.

Layer-Wise Embedding Degradation. To examine how quantization effect propagates through the model's depth, we compute the layer-wise cosine similarity between 16-bit and quantized hidden states at each transformer block, tracing how representational drift accumulates from the input embeddings to the final output layer.

As shown in Fig. 5a in Appendix A, the input embedding layer ($n_{\text{layer}} = 0$) remains universally quantization-invariant across all precision levels. However, as depth increases, encoder-only models exhibit a monotonic, layer-wise degradation under 4-bit quantization. This is most pronounced in DeBERTa-v3-base, where similarity declines steadily from 99.2% at layer 1 to a significant collapse of 65.2% at the final layer. Other encoders, such as DistilBERT-base and MiniLM-L12-v2, follow a pattern of gradual mid-network drift followed by a sharp terminal drop, reaching final similarities of 97.6% and 93.7%, respectively.

Conversely, the layer-wise embedding similarity for decoder-only models as shown in Fig. 5b in Appendix A maintain high fidelity (>99%) through the initial layers, experiencing a "depth-localized" collapse isolated to the terminal 3–5 transformer blocks. LLaMA-3.2-3B sustains 99.4% similarity through layer 24 before declining to 97.6% at the output. SmolLM3-3B holds a similarity score above 99.3% until layer 33, followed by a sharp drop to 91.0% at layer 35.

These contrasting profiles stem from architectural differences in attention masking. Encoder-only models use bidirectional self-attention, so quantization rounding errors compound monotonically as each layer attends over already-degraded representations from all positions. Decoder-only models apply causal masking, which restricts error propagation to the sequential direction and concentrates representational drift in the terminal layers where long-range context is integrated. For clinical deployment under 4-bit constraints, these findings suggest extracting embeddings from intermediate rather than final transformer blocks when using encoder-only models, as final-layer representations show the greatest geometric distortion.

4.6 Topology Fidelity Under Quantization

Embedding similarity characterizes how much individual vectors shift under quantization, but does not capture whether the relational structure of the embedding space is preserved. Specifically, we ask whether health records that are similar under 16-bit remain proximate after compression and whether the global geometry of the space is maintained [36]. To evaluate the fidelity of quantized clinical embedding spaces, we employ two complementary metrics and compare embeddings from quantized variants against the 16-bit embedding baseline.

Centered Kernel Alignment (CKA). We use CKA to measure the global structural similarity between the original and quantized representations. Because CKA is invariant to orthogonal transformations and isotropic scaling, it provides a robust measure of whether the latent space geometry is preserved [37]. As shown in Figs. 6a and 6b in Appendix A, CKA scores remain over 0.99 across early and intermediate layers even under 4-bit quantization, indicating strong preservation of the global manifold. Although terminal-layer degradation appears in certain 4-bit configurations (DeBERTa-v3-base and Qwen2.5-3B), the high aggregate similarity suggests that quantization largely maintains the underlying feature hierarchy and preserves the global geometry of the clinical embedding space.

Nearest Neighbor Topology Preservation (NNTP). We access local relational structure using NNTP [38]. For k = 5 nearest neighbors, NNTP = 1.0 implies that every record's five nearest neighbors are identical across both spaces, providing a direct measure of whether clinically similar patients remain mutually proximate after quantization. As shown in Fig. 7a and 7b in Appendix A, results indicate substantial sensitivity locally. NNTP scores drop sharply after the first hidden layer and stabilize between approximately 0.35 and 0.55 across models. Notably, 4-bit quantization consistently outperforms 8-bit by roughly 10–20% in neighborhood preservation. This pattern suggests that 4-bit quantization-aware optimizations can better maintain local relational proximity than standard 8-bit quantization.

4.7 Memory Footprint Analysis

Decoder-only models at both 4-bit and 8-bit precision achieve an average of 16% reduction in GPU utilization compared to 16-bit precision, whereas encoder-only models exhibit an irregular trend in resource consumption as they shift toward compute bound operations as shown in Figs. 4(a) and (b). This behavior is driven by a spike in GPU memory utilization caused by dequantization overhead. This suggests that aggressive quantization below 8-bit is suited for decoder-only models, while encoder-only models may require quantization-aware training to achieve stable memory efficiency gains [39].

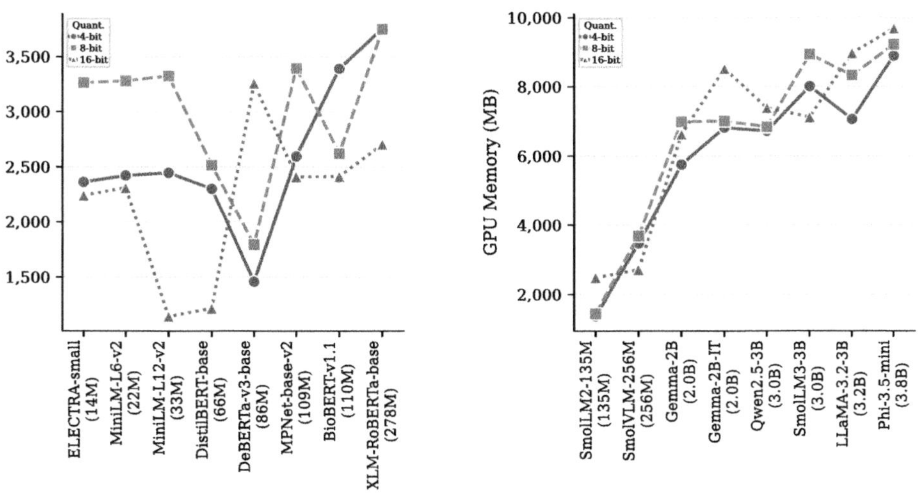

(a) Memory Footprint for Encoder-Only Models

(b) Memory Footprint for Decoder-Only Models

Fig. 4. GPU memory consumption (MB) of encoder-only and decoder-only models under 4-bit, 8-bit, and 16-bit quantization. Models are ordered left to right by increasing parameter count.

5 Conclusion

This study evaluates hyperparameter tuning for the cuML Random Forest Regressor, establishing an optimal configuration that retains 99.3% of the Scikit-learn baseline's predictive accuracy while achieving a 1.5 times reduction in training latency. Building upon this optimized baseline, we demonstrate that the SLM Integrated DMV Framework substantially outperforms the baseline cuML Random Forest Regressor, yielding a 3.07% improvement in predictive accuracy alongside MSE and MAE reductions of 99.43% and 94.47%, respectively. Both encoder-only and decoder-only models achieve statistically equivalent predictive performance, yet encoder-only models require up to 72% less GPU memory. 4-bit quantization preserves over 99.98% of 16-bit embedding baseline while reducing GPU memory by an average of 16%. These findings establish that this quantized SLM embedded framework is a resource-efficient benchmark for AD risk detection, achieving near full-precision performance under computationally constrained clinical environments while also providing a meaningful contribution for future research exploring the use of SLMs in healthcare applications.

Limitations

This study evaluates SLM embeddings on a single static CDC dataset and their generalization to other clinical datasets or electronic health record systems remains unverified. The potential benefits of domain-specific fine-tuning of SLMs on clinical data have not been evaluated in this study.

Future Work

While this study utilizes SLMs in a zero-shot capacity, future work will involve task-specific fine-tuning on privacy-compliant clinical corpora. This will allow the models to better capture subtle linguistic indicators of cognitive decline. By establishing general-purpose SLMs as effective embedding engines within the DMV Framework, this work lays the groundwork for targeted extensions using domain-adapted biomedical models, which may further enhance representational quality for clinical text. Also, the resource efficiency demonstrated in this study paves the way for deploying SLM-based diagnostic tools on edge devices, including clinical tablets and patient smartphones. Future work will focus on enabling fully offline, on-device inference, allowing early detection while preserving patient privacy across a wide range of clinical settings.

Acknowledgments. Computational support for this research was provided by the Modal platform (https://modal.com/). We gratefully acknowledge their support in granting access to NVIDIA A100 GPUs, which enabled the evaluation and validation of our SLM Integrated DMV Framework.

Disclosure of Interests. The authors have no competing interests to declare that are relevant to the content of this article.

Appendix

A Quantization Degradation Analysis

Layer-Wise Embedding Degradation Analysis. As shown in Fig. 5, 8-bit quantization maintains cosine similarity close to 1.0 across all layers, while 4-bit quantization introduces representational degradation in deeper layers.

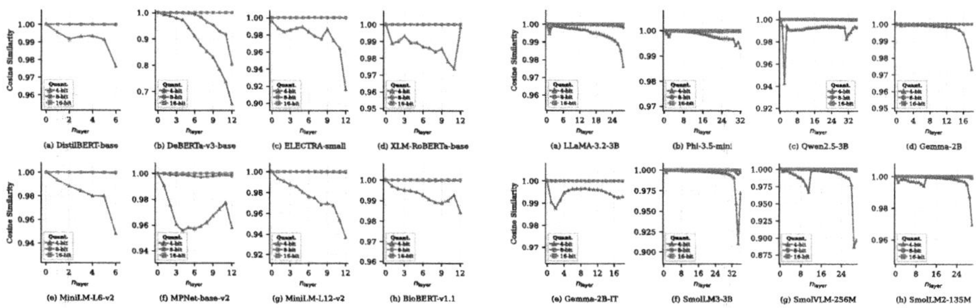

(a) Encoder-Only Models (b) Decoder-Only Models

Fig. 5. Layer-wise cosine similarity between 16-bit and quantized (4-bit, 8-bit) hidden states for encoder-only (a) and decoder-only (b) models, measuring representation fidelity across model depth.

Centered Kernel Alignment (CKA). As shown in Fig. 6, 8-bit quantization preserves global geometric consistency across all layers, while 4-bit quantization introduces structural distortions, with decoder-only models exhibiting greater representational drift in deeper layers than encoder-only models.

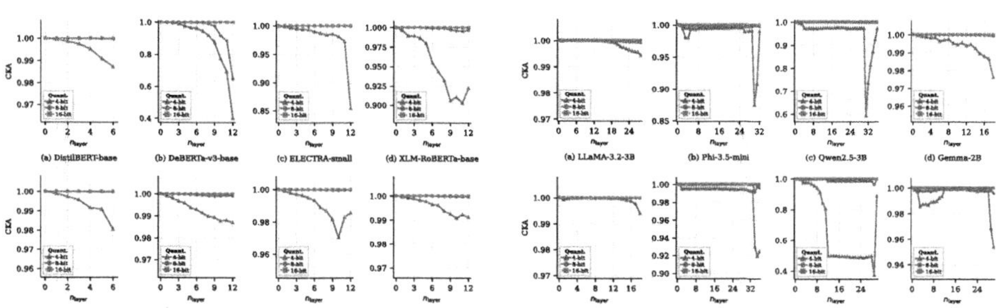

(a) Encoder-Only Models (b) Decoder-Only Models

Fig. 6. Layer-wise CKA between 16-bit and quantized (4-bit, 8-bit) hidden states for encoder-only (a) and decoder-only (b) models, measuring preservation of global embedding geometry across model depth.

Nearest Neighbor Topology Preservation (NNTP). As shown in Fig. 7, quantization triggers a localized collapse of neighborhood structures after the first layer, with 4-bit surprisingly outperforming 8-bit in local topology preservation throughout the network.

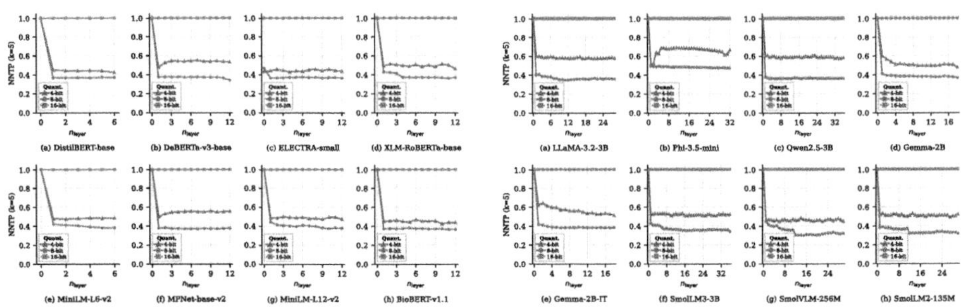

(a) Encoder-Only Models (b) Decoder-Only Models

Fig. 7. Layer-wise NNTP (k=5) between 16-bit and quantized (4-bit, 8-bit) hidden states for encoder-only (a) and decoder-only (b) models, measuring local neighborhood structure retention across model depth.

References

1. Chou, C.-J., et al.: Screening for early Alzheimer's disease: enhancing diagnosis with linguistic features and biomarkers. Front. Aging Neurosci. **16** (2024). https://doi.org/10.3389/fnagi.2024.1451326
2. Agbavor, F., Liang, H.: Predicting dementia from spontaneous speech using large language models. PLOS Dig. Health **1**(12), e0000168 (2022). https://doi.org/10.1371/journal.pdig.0000168
3. Thirunavukarasu, A.J., Ting, D.S.J., Elangovan, K., Gutierrez, L., Tan, T.F., Ting, D.S.W.: Large language models in medicine. Nat. Med. **29**(8), 1930–1940 (2023). https://doi.org/10.1038/s41591-023-02448-8
4. Fu, Y., Peng, H., Ou, L., Sabharwal, A., Khot, T.: Specializing smaller language models towards multi-step reasoning. arXiv preprint arXiv:2301.12726 (2023). https://doi.org/10.48550/arXiv.2301.12726
5. Wang, F., et al.: A comprehensive survey of small language models in the era of large language models: techniques, enhancements, applications, collaboration with LLMs, and trustworthiness. arXiv preprint arXiv:2411.03350 (2024). https://doi.org/10.48550/arXiv.2411.03350
6. Dettmers, T., Zettlemoyer, L.: The case for 4-bit precision: k-bit inference scaling laws. arXiv preprint arXiv:2212.09720 (2022). https://doi.org/10.48550/arXiv.2212.09720
7. Lazarova, S., Grigorova, D., Petrova-Antonova, D.: Detection of Alzheimer's disease using logistic regression and clock drawing errors. Brain Sci. **13**(8), 1139 (2023). https://doi.org/10.3390/brainsci13081139

8. Vichianin, Y., et al.: Accuracy of support-vector machines for diagnosis of Alzheimer's disease, using volume of brain obtained by structural MRI at siriraj hospital. Front. Neurol. **12** (2021). https://doi.org/10.3389/fneur.2021.640696
9. Fraser, K.C., Meltzer, J.A., Rudzicz, F.: Linguistic features identify Alzheimer's disease in narrative speech. J. Alzheimers Dis. **49**(2), 407–422 (2015). https://doi.org/10.3233/jad-150520
10. Velazquez, M., Lee, Y.: Random forest model for feature-based Alzheimer's disease conversion prediction from early mild cognitive impairment subjects. PLoS ONE **16**(4), e0244773 (2021). https://doi.org/10.1371/journal.pone.0244773
11. Hussain, M.Z., et al.: A fine-tuned convolutional neural network model for accurate Alzheimer's disease classification. Sci. Rep. **15**(1) (2025). https://doi.org/10.1038/s41598-025-86635-2
12. Nguyen, M., He, T., An, L., Alexander, D.C., Feng, J., Yeo, B.T.T.: Predicting Alzheimer's disease progression using deep recurrent neural networks. Neuroimage **222**, 117203 (2020). https://doi.org/10.1016/j.neuroimage.2020.117203
13. Priyadharshini, M., Murugesh, V., Rybin, O.: Enhancing Alzheimer's disease classification with a transformer-based model using self-supervised learning. Sci. Rep. **16**(1) (2026). https://doi.org/10.1038/s41598-025-33957-w
14. Zhang, P., Chaudhary, D.: Leveraging clinical record geolocation for improved Alzheimer's disease diagnosis using DMV framework. Biomedicines **13**(10), 2496 (2025). https://doi.org/10.3390/biomedicines13102496
15. Shool, S., Adimi, S., Saboori Amleshi, R., Bitaraf, E., Golpira, R., Tara, M.: A systematic review of large language model (LLM) evaluations in clinical medicine. BMC Med. Inf. Decis. Mak. **25**(1) (2025). https://doi.org/10.1186/s12911-025-02954-4
16. Ghaffari, A., Younesian, S., Nia, V.P., Chen, B., Asgharian, M.: AdpQ: a zero-shot calibration free adaptive post training quantization method for LLMs. arXiv preprint arXiv:2405.13358 (2024). https://doi.org/10.48550/arXiv.2405.13358
17. Frantar, E., Ashkboos, S., Hoefler, T., Alistarh, D.: GPTQ: accurate post-training quantization for generative pre-trained transformers. arXiv preprint arXiv:2210.17323 (2022). https://doi.org/10.48550/arXiv.2210.17323
18. Lin, J., et al.: AWQ: activation-aware weight quantization for LLM compression and acceleration. arXiv preprint arXiv:2306.00978 (2023). https://doi.org/10.48550/arXiv.2306.00978
19. Kogan, A.: Is (Selective) round-to-nearest quantization all you need? arXiv preprint arXiv:2505.15909 (2025). https://doi.org/10.48550/arXiv.2505.15909
20. Devlin, J., Chang, M.-W., Lee, K., Toutanova, K.: BERT: pre-training of deep bidirectional transformers for language understanding. arXiv preprint arXiv:1810.04805 (2018). https://doi.org/10.48550/arXiv.1810.04805
21. Sanh, V., Debut, L., Chaumond, J., Wolf, T.: DistilBERT, a distilled version of BERT: smaller, faster, cheaper and lighter. arXiv preprint arXiv:1910.01108 (2019)
22. Conneau, A., et al.: Unsupervised cross-lingual representation learning at scale. arXiv preprint arXiv:1911.02116 (2019)
23. Song, K., Tan, X., Qin, T., Lu, J., Liu, T.-Y.: MPNet: masked and permuted pre-training for language understanding. arXiv preprint arXiv:2004.09297 (2020). https://doi.org/10.48550/arXiv.2004.09297
24. Touvron, H., et al.: LLaMA: open and efficient foundation language models. arXiv preprint arXiv:2302.13971 (2023). https://doi.org/10.48550/arXiv.2302.13971
25. Abdin, M., et al.: Phi-3 technical report: a highly capable language model locally on your phone. arXiv preprint arXiv:2404.14219 (2024). https://doi.org/10.48550/arXiv.2404.14219

26. Yang, A.: Qwen2 technical report. arXiv preprint arXiv:2407.10671 (2024)
27. Qwen Team: Qwen2.5: a party of foundation models (2024). https://qwenlm.github.io/blog/qwen2.5/
28. Bakouch, E., et al.: SmolLM3: smol, multilingual, long-context reasoner (2025). https://huggingface.co/blog/smollm3
29. Ben Allal, L., et al.: SmolLM2: when smol goes big – data-centric training of a small language model. arXiv preprint arXiv:2502.02737 (2025). https://doi.org/10.48550/arXiv.2502.02737
30. Zhang, P., Chaudhary, D.: Leveraging geolocation in clinical records to improve Alzheimer's disease diagnosis using DMV framework. arXiv preprint arXiv:2502.04288 (2025). https://arxiv.org/abs/2502.04288
31. Draper, N.R., Smith, H.: Applied Regression Analysis. Wiley Series in Probability and Statistics. Wiley (1998). https://doi.org/10.1002/9781118625590
32. Louppe, G., Wehenkel, L., Sutera, A., Geurts, P.: Understanding variable importances in forests of randomized trees. Adv. Neural Inf. Process. Syst. **26** (2013)
33. Hastie, T., Tibshirani, R., Friedman, J.: The Elements of Statistical Learning. Springer Series in Statistics. Springer, New York (2009). https://doi.org/10.1007/978-0-387-84858-7
34. Huang, X., Khetan, A., Cvitkovic, M., Karnin, Z.: TabTransformer: tabular data modeling using contextual embeddings. arXiv preprint arXiv:2012.06678 (2020). https://doi.org/10.48550/arXiv.2012.06678
35. Gong, Y., Liu, L., Yang, M., Bourdev, L.: Compressing deep convolutional networks using vector quantization. arXiv preprint arXiv:1412.6115 (2014). https://doi.org/10.48550/arXiv.1412.6115
36. Indyk, P., Naor, A.: Nearest-neighbor-preserving embeddings. ACM Trans. Algor. **3**(3), 31 (2007). https://doi.org/10.1145/1273340.1273347
37. Kornblith, S., Norouzi, M., Lee, H., Hinton, G.: Similarity of neural network representations revisited. arXiv preprint arXiv:1905.00414 (2019). https://doi.org/10.48550/arXiv.1905.00414
38. Renga Bashyam, K.G., Vadhiyar, S.: Fast scalable approximate nearest neighbor search for high-dimensional data. In: 2020 IEEE International Conference on Cluster Computing (CLUSTER), pp. 294–302. IEEE (2020). https://doi.org/10.1109/cluster49012.2020.00040
39. Li, Z., et al.: SpaceServe: spatial multiplexing of complementary encoders and decoders for multimodal LLMs. In: The Thirty-Ninth Annual Conference on Neural Information Processing Systems (2025). https://openreview.net/forum?id=w4qJ056WhI

Psychology-Grounded Individualized Artificial Mental Model

Prajvi Saxena[1(✉)], Arvind Nagarajan[1], Sabine Janzen[1], and Wolfgang Maass[1,2]

[1] German Research Center for Artificial Intelligence (DFKI), Kaiserslautern, Germany
prajvi.saxena@dfki.de
[2] Saarland University, Saarbrücken, Germany

Abstract. Personalizing rehabilitation requires predicting how individual patients will experience pain yet two patients undergoing identical surgery can report starkly different outcomes. This heterogeneity is not noise: it is systematically driven by personality, beliefs, emotional state, and life circumstances. Psychology has spent decades formalizing precisely these individual differences through validated frameworks such as the Biopsychosocial model and the International Classification of Functioning. Yet existing AI approaches ignore this theoretical structure and treating patient variables as flat numerical features. We propose Psychology-Grounded Artificial Mental Models: a framework that encodes established psychological theory as structured reasoning scaffolds within large language models, enabling them to construct an artificial mental model of each patient before making predictions. To evaluate the framework, we conducted a DiscoverYourself study with 336 participants, and further validated the framework on publicly available PhysioPain dataset (N = 82). Across four LLM families, theory-grounded models achieve up to +56.9% points exact-accuracy improvement over the generic baseline. These results indicate that encoding psychological theory into LLM reasoning substantially improves pain prediction in rehabilitation, with the largest gains observed among older, highly heterogeneous patients where data-driven models are least reliable. The source code and implementation details are available at [1] [1]GitHub:https://github.com/InformationServiceSystems/fedwell-project/tree/main/Individualized_Phase.

Keywords: Artificial Mental Models · Psychology-Grounded AI · Individualized Prediction · Large Language Models · Rehabilitation · Pain Assessment

1 Introduction

Why does the same exercise feel effortless to one patient and agonizing to another even when both underwent the same surgery, share similar demographics, and follow the same rehabilitation protocol? This question lies at the heart of personalized rehabilitation, and its answer is deeply rooted in individual psychology.

S. S. Khan et al. (Eds.): IJCAI 2026, CCIS 3127, pp. 63–77, 2027.
https://doi.org/10.1007/978-981-92-5190-2_5

A patient's experience of pain and exertion is not a straightforward readout of tissue status; it is filtered through layers of personality, emotional regulation, prior beliefs about recovery, lifestyle habits, and social context [6,22]. As populations age and rehabilitation demand grows, the ability to predict these *individual* differences not just population averages becomes increasingly critical for effective, patient-centered care [11,24]. Psychology has investigated this problem for decades producing well-validated theoretical frameworks that explain individual differences in pain and health outcomes. The Biopsychosocial model (BPS) [5] positions every patient at the intersection of biological, psychological, and social forces. The International Classification of Functioning, Disability and Health (ICF) [26] provides a structured taxonomy of body functions, activities, participation, and contextual factors. The Fear-Avoidance model [16] explains how pain-related fear and catastrophizing drive disability beyond what injury alone predicts. Cognitive-Behavioral frameworks [3] identify how maladaptive thought patterns sustain pain perception long after tissue healing. Together, these frameworks are not abstract constructs, they are the cognitive tools experienced clinicians use daily to reason about patients as whole persons.

Yet a striking gap exists: while AI systems for healthcare have grown increasingly powerful, they have largely ignored the individualized patients characteristics grounded in psychological theory. Current applications of Large Language Models (LLMs) to clinical prediction rely either on generic prompting of feeding patient data without any structured reasoning guidance or on domain-specific fine-tuning, which may capture statistical patterns but might fails to reasons about patients in clinically meaningful ways [2,20]. Neither approach gives the model what a skilled clinician possesses: a *theory of the individual:* a structured, theoretically grounded account of how biological, psychological, and social factors interact within a specific individual to produce a specific pain response under specific conditions.

In this paper, we examine whether encoding validated psychological theory directly into LLM reasoning transforms clinical prediction from population-level pattern matching into individualized inference grounded in a structured understanding of the patient. We introduce *Psychology-Grounded Artificial Mental Models* a framework that encodes established psychological theories of individual differences directly into LLM reasoning, enabling it to construct an individualized mental model of each patient before predicting perceived pain on a (1–5) Numeric Rating Scale (NRS) during knee and lower-limb rehabilitation exercises. Given a structured patient profile and exercise context, the Biopsychosocial model and ICF framework are operationalized as a structured prompting architecture, instructing the LLM to integrate physical capacity, psychological disposition, lifestyle, and functional context into a holistic patient persona from which predictions are derived.

The distinction has direct predictive consequences. A generic LLM assigns a pain score to a 55-year old patient with high anxiety, poor sleep, and low emotional stability based on surface correlations. Our framework reasons over the patient persona: anxiety amplifies pain perception, poor sleep lowers pain thresholds, and low emotional stability predicts catastrophizing, arriving at substantially more individualized prediction. This paper makes three contributions:

1. **A theoretically grounded reasoning framework for clinical prediction:** We demonstrate that established psychological theories of individual differences, including the Biopsychosocial model and the ICF framework, can be operationalized as structured LLM reasoning scaffolds, creating what we term *Psychology-Grounded Artificial Mental Models*. This bridges clinical psychology theory and AI-driven prediction.
2. **Systematic empirical evidence across LLMs and prompts:** Through experiments across four LLM families (LLaMA 3.1, Qwen 2.5, Mistral, and Phi 3) and multiple prompting strategies (zero-shot baseline, in-context learning, BPS-only, ICF-only, BPS+ICF, and exploratory pain-theory variants), we show that psychology-grounded prompting improves individualized prediction over generic baselines.
3. **Validation across two rehabilitation-related datasets:** We evaluate the framework on the DiscoverYourself study with 336 participants and further validate it on the publicly available PhysioPain dataset ($N = 82$ after pre-processing). The BPS+ICF framework achieves consistent improvements across model families, with gains of up to 56.9% points in exact accuracy over the generic baseline on the DiscoverYourself dataset.

1.1 Background and Related Work

The Biopsychosocial (BPS) model [5] establishes that health outcomes emerge from the dynamic interplay of biological, psychological, and social factors rather than biological pathology alone. In rehabilitation, this matters profoundly: two patients with identical knee surgeries can follow entirely different recovery trajectories depending on fear of movement, personality-driven coping strategies, and available social support [15]. The ICF [26] complements the BPS model by providing a structured taxonomy of functioning across body functions, activities, participation, and contextual factors. Together, these frameworks give clinicians a principled vocabulary for reasoning about individual patients; the BPS model supplies the causal structure, the ICF supplies the descriptive completeness.

Johnson-Laird's theory of mental models [13] proposes that humans reason not through formal logic but by constructing internal representations of situations that capture the relationships between entities and their properties. Applied to clinical reasoning, an experienced clinician builds a mental model of each patient, an integrated, dynamic representation of who this person is, how they are likely to respond, and why. Our framework operationalizes this concept computationally: we construct Artificial Mental Models (AMMs) by encoding psychological theory as structured reasoning scaffolds within LLMs, enabling them to build an analogous patient representation before making predictions [10,17].

LLMs have demonstrated strong capabilities in clinical text analysis, medical question answering, and patient simulation [4,20,25]. Yet common paradigms for clinical prediction, such as generic prompting or domain-specific adaptation, may fall short for individualized reasoning. Generic prompting provides no structured guidance for integrating heterogeneous patient factors. Domain-specific adaptation on small clinical datasets can risk capturing surface correlations rather than clinically meaningful patterns, and produces reasoning that is neither interpretable nor verifiable against clinical theory [2]. Our framework addresses this gap directly by encoding established psychological theory as explicit reasoning scaffolds, giving LLMs the same structured understanding of individual differences that clinicians rely on.

2 Psychology-Grounded AMM Framework

We formulate individualized pain prediction as an ordinal estimation task. Given a structured participant profile x_i and an exercise context e, the objective is to predict the participant's self-reported perceived difficulty $\hat{y}_i$ on a five-point Numeric Rating Scale (NRS):

$$f(x_i, e) = \hat{y}_i, \qquad \hat{y}_i \in \{1, 2, 3, 4, 5\} \tag{1}$$

where 1 denotes *not difficult at all* and 5 denotes *extremely difficult*. Critically, the target is a subjective self-report; the model must capture how an individual *experiences* an exercise, not merely whether they can physically perform it.

Our framework operationalizes this through three sequential components: (i) Psychology-driven input structuring, (ii) persona construction, and (iii) theory-guided prediction. Figure 1 illustrates the overall architecture.

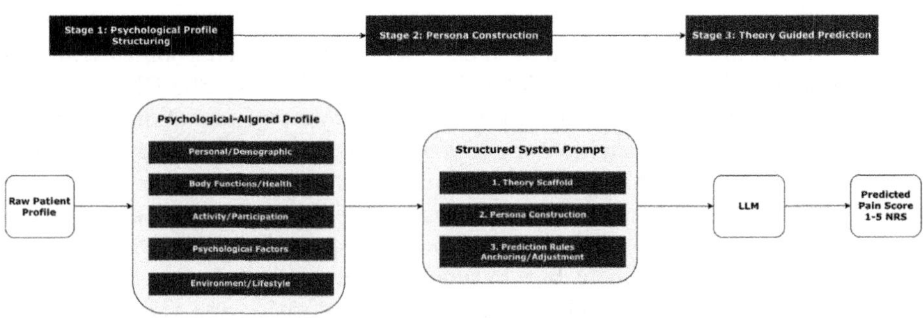

Fig. 1. Psychology-grounded AMM framework for pain-score prediction: Raw patient profiles are organized into theory-aligned biopsychosocial sections, converted into a structured system prompt with persona and prediction constraints, and passed to an LLM for scoring on a 1–5 NRS pain scale.

2.1 Psychology-Driven Input Structuring

A standard approach to clinical prediction treats patient attributes as a flat feature vector and assign all reasoning about interactions and context to the model. Our framework instead imposes the structure of the BPS model and the ICF framework on the input itself, so that the LLM encounters patient information already organized according to established clinical reasoning categories. Concretely, each patient profile is partitioned into five theory-aligned sections that jointly cover the BPS and ICF dimensions:

1. **Personal/Demographic Factors** (ICF: Personal Factors): Age, gender, employment status, industry, work type.
2. **Body Functions/Health Condition** (BPS: Biological; ICF: Body Functions & Structures): Disability status, sleep hours, sleep problems, overall health, mobility, surgery history, surgical complications, recovery period, physical therapy history and adherence.
3. **Activity/Participation** (ICF: Activities & Participation): Daily activities performed without difficulty, ability to perform the target exercise (e.g., 10 squats or sit-to-stand), maximum repetitions achievable.
4. **Psychological Factors** (BPS: Psychological; ICF: Personal Factors): Current mood, stress or anxiety in the past 24 h, emotional state, Big Five personality traits[1], daily-life stress level.
5. **Environmental/Lifestyle Factors** (BPS: Social; ICF: Environmental Factors): Types of physical activities practiced, exercise frequency (days per week), session duration, mood–exercise relationship (whether the participant exercises more when in a positive or negative mood).

2.2 Persona Construction

The system prompt instructs the LLM to process the structured patient profile based on mental model theory [9,13] and construct the internal representation persona of the patient. Rather than mapping features directly to a prediction, the model must first internally construct a realistic behavioral persona of the participant. The persona integrates eight dimensions:

- Physical capacity and mobility limitations
- Health conditions and recovery status
- Confidence or caution tendencies
- Current emotional state and stress exposure
- Exercise habits and physical activity level
- Personality-driven self-rating tendencies
- Whether the participant typically rates difficulty as low, moderate, or high
- Likely gap between perceived and actual difficulty

[1] Assessed via the Ten-Item Personality Inventory (TIPI) [7]: extraversion, agreeableness, conscientiousness, emotional stability, openness to experience, plus their reverse-scored counterparts.

2.3 Theory-Guided Prediction

The structured profile and the constructed persona together form the complete input to the LLM. Rather than presenting raw patient attributes directly to the model, the framework ensures that the model receives an psychological-interpreted representation. The prediction is derived from this representation, not from the raw features. The system prompt encodes the reasoning process in three stages. First, the model processes the theory-structured input and constructs the internal persona as described in Sect. 2.2. Second, guided by the encoded psychological frameworks, the model reasons over the persona to assess how the individual's biological, psychological, and social factors are likely to interact under the given exercise context. Third, the model produces a score on the target ordinal scale, where the prediction reflects the individual's expected subjective experience rather than an objective physical capacity estimate. This design ensures that prediction is grounded in a theoretically coherent account of the individual.

2.4 Prediction Mechanism

All LLM-based configurations use direct model generation followed by rule-based score extraction. Given a structured prompt p, the model generates a short textual response:

$$r = \mathrm{LLM}(p) \tag{2}$$

The generated response r is then parsed to obtain the predicted score $\hat{y} \in \{1, \ldots, 5\}$. The prompt explicitly instructs the model to output only one score from 1 to 5 and not to include reasoning. After generation, the response is normalized and checked for a valid ordinal score. If a valid score is found, it is used as the prediction; otherwise, the output is treated as invalid according to the evaluation protocol.

Using the same generation-and-parsing procedure across all configurations ensures that comparisons between prompting strategies reflect differences in the reasoning scaffold rather than differences in the prediction mechanism. All configurations therefore produce predictions on the same discrete 1–5 scale.

3 Data

We evaluate our framework on two independent datasets differing in provenance, population, and target outcome. Full dataset details are provided in Appendix C.

3.1 DiscoverYourself Dataset

The primary dataset comprises N=336 participant profiles collected across two waves of the DiscoverYourself study. In both waves, participants completed a structured digital questionnaire and performed ten bodyweight squats guided by video instruction. Participants self-reported perceived difficulty before and after the exercise on a 1–5 Numeric Rating Scale (NRS), where 1 indicates *not difficult at all* and 5 indicates *extremely difficult* [18]. The post-exercise NRS score is the prediction target. The questionnaire captures variables spanning all BPS and ICF dimensions as described in Sect. 2.1: demographics, health and body functions, activity and participation, psychological state, and lifestyle and environmental factors. Variables are expressed as natural-language text rather than numeric codes, enabling the LLM to leverage the semantic content of responses during reasoning. The sample skews young and physically active, which we discuss as a limitation in Sect. 5.

3.2 PhysioPain Dataset

The PhysioPain dataset [23] comprises $N = 82$ participant profiles after preprocessing. The original survey contains a larger set of responses ($N = 99$), but we retain only records with a valid response for the target item, "Rate the severity of your pain (Likert scale)", corresponding to column AH in the source file. Records without this target value are omitted from evaluation. The dataset is a publicly available pain assessment survey covering demographics, health status, psychological state, and lifestyle factors. The prediction target is a self-reported pain outcome on the same 1–5 NRS scale. This dataset serves as an external validation set: collected through a different modality, from a partially different population, and targeting pain perception rather than exercise difficulty, it tests whether the framework generalizes beyond the primary study setting.

4 Experiments and Results

4.1 Experimental Setup

We evaluate four instruction-tuned LLM families: LLaMA 3.1 8B Instruct [8], Qwen 2.5 7B Instruct [21], Mistral 7B Instruct [12], and Phi-3 Mini Instruct [1]. Each model is evaluated under five primary configurations: a generic baseline, In-Context Learning (**ICL**), where 10 labeled examples for the DiscoverYourself dataset and 5 for the PhysioPain dataset are provided; **BPS-only**, where the system prompt encodes the BPS framework; **ICF-only**, where the prompt is structured around functional health categories; and **BPS+ICF**, the full framework with combined theoretical scaffolding, persona construction, and prediction rules. We additionally evaluate three exploratory pain-theory prompt variants: Fear-Avoidance, Cognitive-Behavioral, and Gate Control Theory prompting.

4.2 Results on DiscoverYourself Dataset

Table 1 reports exact accuracy across all model family on the DiscoverYourself dataset ($N = 336$).

The BPS+ICF configuration produces the strongest result for all four model families, with exact accuracy ranging from 62.8% to 63.7%. This consistency

Table 1. Exact accuracy (%) on the DiscoverYourself dataset (N=336). Best result per row in **bold**.

Model	Baseline	ICL(10-shots)	BPS-only	ICF-only	BPS+ICF
LLaMA 3.1 8B	7.4	19.6	54.8	29.5	**63.7**
Phi-3 Mini 4K	6.6	27.9	32.5	9.2	**63.5**
Mistral 7B v0.3	7.4	22.3	31.9	27.4	**62.8**
Qwen 2.5 7B	8.0	38.6	7.0	19.7	**63.4**

across LLaMA 3.1, Phi-3, Mistral, and Qwen 2.5 suggests that the gain is not tied to a single architecture. In particular, Qwen 2.5 improves substantially under the full BPS+ICF configuration, indicating that the combined theoretical scaffold is more stable than partial theory-only prompting for this dataset.

4.3 Results on PhysioPain Dataset

Table 2 reports exact accuracy on the PhysioPain dataset ($N = 82$) targeting pain perception rather than exercise difficulty.

Table 2. Exact accuracy (%) on the external pain dataset (N=82). Best result per row in **bold**.

Model	Baseline	ICL(5-shots)	BPS-only	ICF-only	BPS+ICF
LLaMA 3.1 8B	39.0	35.0	25.6	25.7	**58.5**
Qwen 2.5 7B	31.7	37.6	50.0	23.6	**53.7**
Phi-3 Mini 4K	43.9	31.1	38.0	26.3	**58.5**
Mistral 7B v0.3	35.4	34.0	37.3	21.1	**54.9**

The framework also generalizes to the external PhysioPain dataset, with BPS+ICF achieving the strongest result for all four model families. LLaMA 3.1 and Phi-3 both reach 58.5% exact accuracy, followed by Mistral at 54.9% and Qwen 2.5 at 53.7%. Although the dataset is smaller and targets pain severity rather than post-exercise difficulty, the consistent performance of BPS+ICF suggests that the combined theoretical scaffold transfers beyond the primary 10-squat setting.

4.4 Ablation: Contribution of Psychological Theory

To quantify the contribution of psychological grounding, we compare the generic baseline against the full BPS+ICF framework on the DiscoverYourself dataset. The baseline represents prediction without the full psychology-grounded reasoning scaffold, while BPS+ICF adds theory-driven input structuring, persona construction, and combined biopsychosocial and functional reasoning. Figure 2 reports exact accuracy for the four base LLMs. Across all four model

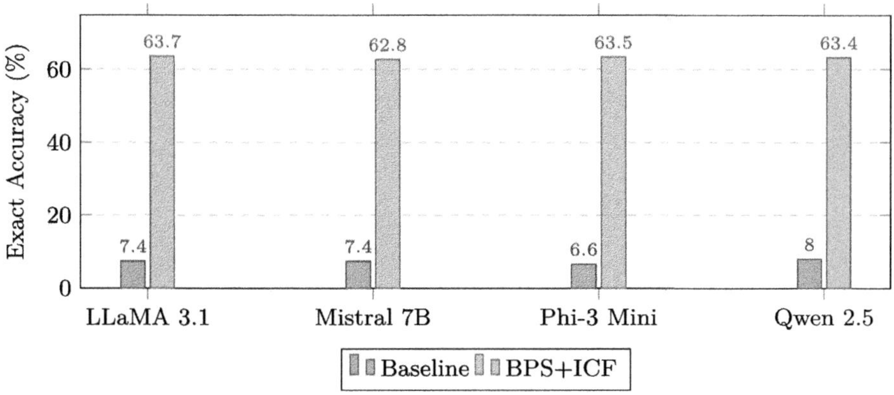

Fig. 2. Exact accuracy of the baseline and BPS+ICF framework on the DiscoverYourself dataset. BPS+ICF improves across all models, with gains of +55.4 to +56.9 percentage points.

families, BPS+ICF substantially outperforms the baseline. LLaMA 3.1 improves from 7.4% to 63.7% exact accuracy (+56.3 pp), Mistral 7B improves from 7.4% to 62.8% (+55.4 pp), Phi-3 Mini improves from 6.6% to 63.5% (+56.9 pp), and Qwen 2.5 improves from 8.0% to 63.4% (+55.4 pp). These results show that the gain is consistent across model families and is not tied to a single architecture. The consistent improvement over the baseline indicates that theory-grounded prompting changes how the models use participant information. Rather than treating survey fields as isolated variables, the BPS+ICF framework encourages the model to reason over the interaction between biological condition, psychological state, lifestyle context, activity limitations, and participation-level factors. This supports the central claim that individualized rehabilitation prediction benefits from explicitly encoding psychological and functional theory into the LLM inference process.

4.5 Exploratory Pain-Theory Prompt Variants

To verify that the final BPS+ICF configuration was not selected arbitrarily, we additionally evaluated three pain-theory prompt variants: Fear-Avoidance, Cognitive-Behavioral, and Gate Control Theory prompting. These variants are treated as exploratory alternatives rather than as the main proposed framework. Unlike BPS+ICF, which combines biological, psychological, social, functional, participation-level, and contextual factors, these theories focus on narrower mechanisms of pain perception, avoidance, coping, and sensory modulation. Table 3 reports the results on the DiscoverYourself dataset, while Table 4 reports the corresponding results for the PhysioPain dataset.

Table 3. Exploratory exact accuracy (%) of pain-theory prompt variants on the DiscoverYourself dataset (N = 336).

Model	Fear-Avoidance	Cognitive-Behavior	Gate Control
LLaMA 3.1 8B	4.5	3.9	4.5
Phi-3 Mini 4K	22.6	12.2	14.0
Mistral 7B v0.3	15.2	10.4	11.9
Qwen 2.5 7B	48.5	7.7	8.6

Table 4. Exploratory exact accuracy (%) of pain-theory prompt variants on the PhysioPain dataset (N=82).

Model	Fear-Avoidance	Cognitive-Behavior	Gate Control
LLaMA 3.1 8B	18.3	18.3	18.3
Phi-3 Mini 4K	31.5	53.7	56.1
Mistral 7B v0.3	23.2	30.5	25.6
Qwen 2.5 7B	50.0	43.9	32.9

5 Discussion

Why Theory-Grounded Prompting Improves Prediction: The main finding is that BPS+ICF prompting consistently improves over the generic baseline across all four model families on the DiscoverYourself dataset. We attribute this improvement to the structure imposed by the psychological reasoning scaffold. Instead of treating participant variables as isolated fields, the prompt organizes them into clinically meaningful biological, psychological, social, functional, and contextual dimensions. This gives the model an explicit framework for interpreting heterogeneous participant information before producing a prediction.

Implications for Aging and Rehabilitation: Aging populations present two interrelated challenges: data scarcity and high individual heterogeneity [14]. Our framework addresses both by encoding established psychological theory into the prompt, providing domain knowledge that would otherwise require large-scale training data to approximate. Rather than learning from thousands of patient profiles that pain, reduced mobility, psychological state, and social context interact, the model is explicitly guided to reason through these relationships using BPS and ICF principles. Constructing an individualized persona per patient also makes the framework suitable for heterogeneous rehabilitation settings, where patients with similar diagnoses may differ substantially in psychological disposition, lifestyle, and functional capacity.

Limitations: The DiscoverYourself dataset skews young and physically active, limiting direct generalizability to elderly rehabilitation cohorts; targeted evaluation with older adults remains necessary. The PhysioPain dataset is also small after preprocessing, with 82 usable records, so external validation results should be interpreted cautiously. The 1–5 NRS target is subjective and coarse, and exact-match accuracy may not fully capture clinical utility. All evaluations are cross-sectional; whether psychology-grounded reasoning maintains its advantage across longitudinal rehabilitation trajectories remains an open question. Finally, although BPS+ICF performs consistently across the evaluated model families, the exploratory pain-theory variants show that narrower theoretical prompts can behave differently across architectures and datasets.

6 Conclusion

We presented a framework that encodes established psychological theory as structured reasoning scaffolds within LLMs, enabling individualized prediction of self-reported exercise difficulty and pain-related outcomes. Evaluated across four LLM families and two datasets, the BPS+ICF framework produces consistent improvements over the generic baseline, with gains of up to 56.9 percentage points on the DiscoverYourself dataset. The central finding is that individualized rehabilitation prediction benefits not only from model scale or data availability, but also from theory: providing LLMs with clinically grounded frameworks such as the Biopsychosocial model and ICF improves how they interpret heterogeneous participant profiles. Future work will extend evaluation to older clinical cohorts, incorporate longitudinal rehabilitation trajectories, and examine additional psychological theories of pain and recovery.

Ethics and Broader Impact Statement

Study Ethics and Data Collection. The DiscoverYourself dataset was collected under the oversight of the research institution's ethical review procedures. All participants provided informed consent prior to enrollment and were informed that their self-reported data would be used for AI model development. Participation was voluntary, with no clinical intervention administered and no therapeutic decisions influenced by the study. All participant records were pseudonymized prior to analysis, with identifying information removed from the dataset. The PhysioPain dataset [23] is publicly available under the CC BY 4.0 license[2].

Ethical Considerations of Psychological Profiling. A distinctive feature of our approach and one that requires explicit ethical reflection is that it encodes established psychological theories into an AI system's reasoning process, instructing LLMs to construct holistic behavioral personas of individual patients. While this is precisely what makes the approach effective, it also means that the system engages in structured psychological profiling, integrating personality traits,

[2] https://creativecommons.org/licenses/by/4.0/.

emotional states, coping tendencies, and lifestyle patterns into an internal model of "who this patient is." We emphasize three safeguards. First, the constructed persona is an *internal reasoning scaffold*, not a stored or communicated artifact: it exists only within the model's inference process and is not saved, shared, or made accessible outside the prediction pipeline. Second, the reasoning is grounded in *established, peer-reviewed clinical frameworks* (BPS model, ICF) rather than ad-hoc or algorithmically discovered categories, ensuring that the profiling aligns with accepted clinical practice. Third, the persona is constructed from *self-reported* data that participants voluntarily provided, not from inferred or behavioral data collected without their knowledge. Despite these safeguards, we acknowledge that computational psychological profiling carries inherent risks. We strongly advocate that any deployment of psychology-grounded AMMs be subject to institutional ethics review, with clear governance over data access, model outputs, and downstream use.

Fairness and Representational Limitations. The DiscoverYourself dataset skews toward young, physically active participants, which limits the representativeness of our evaluation for the elderly populations most relevant to clinical deployment. The psychological frameworks we employ (BPS model, ICF, TIPI personality assessment) were developed and validated primarily in Western clinical and research contexts; their applicability across diverse cultural, linguistic, and socioeconomic backgrounds should not be assumed without further validation. We did not conduct formal fairness audits (e.g., across age, gender, or ethnicity subgroups) due to insufficient subgroup sample sizes for reliable comparison, and we flag this as a necessary step before any clinical deployment.

Reproducibility. All experiments use publicly available, instruction-tuned LLMs (LLaMA 3.1 8B, Qwen 2.5 7B, Mistral 7B v0.3, Phi-3 Mini 4K) accessed through standard inference APIs. The system prompts encoding the BPS and ICF frameworks are described in detail in Appendix A; exact prompt templates and code are available at[3].

Acknowledgments. This work was partially funded by the German Federal Ministry of Education and Research (BMBF) under the contract 01IW23004.

A Biopsychosocial–ICF Prompt Template

This appendix provides a condensed version of the BPS–ICF system prompt. Participant-specific survey fields were inserted separately during inference.

```
You are an expert in human behavior modeling and functional health interpretation.

Given participant survey data from an exercise study, predict the participant's
self-reported ACTUAL difficulty after performing 10 squats.
```

[3] GitHub:https://github.com/InformationServiceSystems/fedwell-project/tree/main/Individualized_Phase.

```
Internally construct a behavioral persona rather than treating fields as isolated
variables. Consider physical capacity, health and mobility limitations, emotional
state, stress or anxiety, exercise habits, confidence, and likely self-rating style.

Use perceived difficulty before exercise as the default prediction. Change the
score only when the full profile gives clear evidence that actual difficulty
would be higher or lower. If uncertain, stay close to perceived difficulty.

Use the combined Biopsychosocial and ICF frameworks internally:
- BPS: biological, psychological, and social/lifestyle factors.
- ICF: body functions, activities, participation, and contextual factors.

Return only one score from 1 to 5.
1 = not difficult at all; 2 = slightly difficult; 3 = moderately difficult;
4 = very difficult; 5 = extremely difficult.
```

B Theoretical Descriptions

Biopsychosocial Model. The biopsychosocial model explains health and illness as interactions between biological, psychological, and social factors rather than as purely biomedical conditions [5]. In this study, it structures prompts around physical health, emotional state, confidence, lifestyle, and activity habits.

ICF. The International Classification of Functioning, Disability and Health describes functioning through body functions and structures, activities, participation, and contextual factors [26]. We use it to frame the 10-squat task as an activity-level outcome influenced by functional capacity and individual context.

C Dataset Descriptions

C.1 DiscoverYourself Dataset

The DiscoverYourself dataset contains 336 participant records from exercise assessments involving the 10-squat task. Although collected across multiple sessions or sources, the records are treated as one combined rehabilitation dataset. Each profile includes demographic, health, psychological, lifestyle, and task-specific variables. The prediction target is the participant's self-reported actual difficulty after performing 10 squats, represented on a 1–5 ordinal scale. Perceived difficulty before exercise is included as an input feature, while post-exercise actual difficulty is used as the ground-truth label.

C.2 PhysioPain Dataset

The PhysioPain dataset is used as an external validation dataset beyond the 10-squat setting. After excluding records without a valid target response, 82 participant profiles remain. The dataset contains demographic, health, pain-related, psychological, and lifestyle variables. The target is a self-reported pain-severity outcome mapped to a 1–5 ordinal scale. This tests whether the same Biopsychosocial–ICF prompting framework can transfer from exercise-difficulty prediction to pain-severity prediction.

D Implementation Details

All experiments were implemented in Python using PyTorch and the Hugging Face `transformers` library. Models were loaded with `AutoTokenizer` and `AutoModelForCausalLM`, using `bfloat16` precision and automatic device placement via `device_map="auto"` [19]. If the tokenizer did not define a padding token, the end-of-sequence token was used as the padding token. The maximum input length was set to 4096 tokens.

Ground-truth and perceived-difficulty labels were parsed from the dataset using a rule-based extraction function. Exact accuracy was computed over records with a valid ground-truth label. Records without a valid target label were excluded from metric calculation.

References

1. Abdin, M., Aneja, J., Awadalla, H., et al.: Phi-3 technical report: a highly capable language model locally on your phone (2024). https://arxiv.org/abs/2404.14219
2. Barnett, S., Brannelly, Z., Kurniawan, S., Wong, S.: Fine-tuning or fine-failing? Debunking performance myths in large language models. arXiv preprint arXiv:2406.11201 (2024)
3. Beck, A.T., Rush, A.J., Shaw, B.F., Emery, G., DeRubeis, R.J., Hollon, S.D.: Cognitive therapy of depression. Guilford Publications (2024)
4. Cao, W., Qu, M., Zhu, T., et al.: Benchmarking large language models against human experts in rehabilitation medicine: a multidimensional evaluation. J. Neuroeng. Rehabil. **23**, 84 (2026). https://doi.org/10.1186/s12984-026-01903-0
5. Engel, G.L.: The need for a new medical model: a challenge for biomedicine. Science **196**(4286), 129–136 (1977). https://doi.org/10.1126/science.847460
6. Gatchel, R.J., Peng, Y.B., Peters, M.L., Fuchs, P.N., Turk, D.C.: The biopsychosocial approach to chronic pain: scientific advances and future directions. Psychol. Bull. **133** (2007). https://doi.org/10.1037/0033-2909.133.4.581
7. Gosling, S.D., Rentfrow, P.J., Swann, W.B., Jr.: A very brief measure of the big-five personality domains. J. Res. Pers. **37**(6), 504–528 (2003)
8. Grattafiori, A., et al.: The llama 3 herd of models. arXiv preprint arXiv:2407.21783 (2024)
9. Janzen, S., Saxena, P., Agnes, C., et al.: Ki in der rehabilitation - anwendung künstlicher mentaler modelle für eine personalisierte medizin. Bundesgesundheitsblatt - Gesundheitsforschung - Gesundheitsschutz **68**, 889–897 (2025). https://doi.org/10.1007/s00103-025-04090-w
10. Janzen, S., Maass, W., Saxena, P.: Investigation of artificial mental models for healthcare AI systems (2024)
11. Janzen, S., et al.: Fedwell: life-long federated user and mental modeling for health and well-being. In: RPEatCAiSE25: Research Projects Exhibition at the International Conference on Advanced Information Systems Engineering, Vienna, Austria, 16–20 June 2025. International Conference on Advanced Information Systems Engineering (CAiSE-2025) (2025)
12. Jiang, A.Q., et al.: Mistral 7b (2023). https://arxiv.org/abs/2310.06825
13. Johnson-Laird, P.N.: Mental models (1989)

14. Khan, H.T.A., Addo, K.M., Findlay, H.: Public health challenges and responses to the growing ageing populations. Public Health Chall. **3**(3), e213 (2024). https://doi.org/10.1002/puh2.213
15. Kunnen, M., Dionigi, R.A., Litchfield, C., Moreland, A.: Psychological barriers negotiated by athletes returning to soccer (football) after anterior cruciate ligament reconstructive surgery. Ann. Leisure Res. **26**(4), 545–566 (2023)
16. Leeuw, M., Goossens, M.E.J.B., Linton, S.J., Crombez, G., Boersma, K., Vlaeyen, J.W.S.: The fear-avoidance model of musculoskeletal pain: current state of scientific evidence. J. Behav. Med. **30**(1), 77–94 (2007). https://doi.org/10.1007/s10865-006-9085-0
17. Maass, W., Janzen, S., Saxena, P., Mukherjee, S.: Evolving afferent architectures: biologically-inspired models for damage-avoidance learning (2026). https://arxiv.org/abs/2602.04807
18. Nugent, S.M., Lovejoy, T.I., Shull, S., Dobscha, S.K., Morasco, B.J.: Associations of pain numeric rating scale scores collected during usual care with research administered patient reported pain outcomes. Pain Med. **22**(10), 2235–2241 (2021)
19. Saxena, P., Janzen, S., Maaß, W.: Streamlining LLMs: adaptive knowledge distillation for tailored language models. In: Ebrahimi, A., Haider, S., Liu, E., Haider, S., Leonor Pacheco, M., Wein, S. (eds.) Proceedings of the 2025 Conference of the Nations of the Americas Chapter of the Association for Computational Linguistics: Human Language Technologies (Volume 4: Student Research Workshop), pp. 448–455. Association for Computational Linguistics, Albuquerque (2025). https://doi.org/10.18653/v1/2025.naacl-srw.43. https://aclanthology.org/2025.naacl-srw.43/
20. Singhal, K., et al.: Large language models encode clinical knowledge. Nature **620**(7972), 172–180 (2023). https://doi.org/10.1038/s41586-023-06291-2
21. Team, Q.: Qwen2.5: a party of foundation models (2024). https://qwenlm.github.io/blog/qwen2.5/
22. Timm, A., Schmidt-Wilcke, T., Blenk, S., Studer, B.: Altered social decision making in patients with chronic pain. Psychol. Med. **53**(6), 2466–2475 (2023)
23. Toktay, B., İkbal Işık Orhan, Yıldırım, E., Akbulut, F.P., Catal, C.: Multimodal insights into diverse pain experiences: physiopain dataset. Data Brief **62**, 111992 (2025). https://doi.org/10.1016/j.dib.2025.111992. https://www.sciencedirect.com/science/article/pii/S2352340925007164
24. Wade, D.T., Halligan, P.W.: The biopsychosocial model of illness: a model whose time has come. Clin. Rehabil. **31**(8), 995–1004 (2017). https://doi.org/10.1177/0269215517709890
25. Wang, R., et al.: Patient-{\Psi}: using large language models to simulate patients for training mental health professionals. arXiv preprint arXiv:2405.19660 (2024)
26. World Health Organization: International Classification of Functioning, Disability and Health: ICF. World Health Organization, Geneva (2001). https://iris.who.int/handle/10665/42407

CoRDial: A Multi-agent Conversational Framework for Care Robotics

Vihanga Ashinsana Wijayasekara(✉), Nirmalie Wiratunga, Pedram Salimi, Ikechukwu Nkisi-Orji, David Corsar, Kay Cooper, Erin Hart-Winks, Lyndsay Alexander, and Kyle Martin

Robert Gordon University, Aberdeen, UK
{v.wijayasekara,n.wiratunga,p.salimi1,i.nkisi-orji,d.corsar1,k.cooper,
l.a.alexander,k.martin3}@rgu.ac.uk

Abstract. As embodied agents move into everyday assistive settings, they must combine practical task support with socially appropriate interaction while adapting to individual user abilities, safety needs, preferences, and contextual constraints. However, monolithic conversational architectures, in which a single model is responsible for interpreting, managing, and responding to dialogue, often struggle with multi-intent utterances and personalisation. In response, we present CoRDial, a multi-agent conversational framework for care robotics, co-designed through seven workshops with 27 older adults and caregivers. The architecture classifies conversational tasks to specialised agents, coordinated by a class prototype-based routing mechanism, ProtoKNN Router, for low latency intent classification. Dedicated safety and explainability agents, including an Ethics Check LLM and a counterfactual explainer, support dynamic consent by helping users understand, question, or revise robot actions when requests cannot be fulfilled, or can only be partially fulfilled, due to safety, ethical, or contextual constraints. Evaluations show that our ProtoKNN achieves accuracy ranging from 70.67% (1-shot) to 96.00% (10-shot, Mahalanobis), offering a strong latency-accuracy trade-off well-suited to scalable elder care deployments.

Keywords: Agentic AI · Care Robotics · Personalised HRI

1 Introduction

As populations age, the need for thoughtful, person centred support at home is becoming increasingly urgent. Conversational and robotic systems offer promising ways to support everyday activities and reduce social isolation among older adults, but they continue to face important challenges in reliability, safety, and adaptation to each person's needs, abilities, and preferences.

In this paper, we introduce CoRDial, a modular multi-agent conversational framework for elder-care robots. Rather than relying on a single, monolithic AI system, CoRDial routes user requests to specialised agents for different kinds of

S. S. Khan et al. (Eds.): IJCAI 2026, CCIS 3127, pp. 78–90, 2027.
https://doi.org/10.1007/978-981-92-5190-2_6

support, such as medication, reminders, practical assistance, and casual conversation. This structure allows the system to better handle complex or multi-intent requests, personalise responses, and shift tone appropriately between social and task-focused interactions. CoRDial is underpinned by participatory co-design. The framework uses counterfactual explanations in a novel way: to communicate risks, support dynamic consent, and promote safer interactions by clarifying when, why, or under what conditions a user request cannot be fulfilled, or can only be partially fulfilled.

2 Related Work

This section reviews three strands of prior work relevant to CoRDial: co-design in elder care technologies, conversational AI for older adults, and simulation platforms for embodied agents.

2.1 User Centered Development Approach

User-centred design (UCD) and co-design have become central approaches for developing AI systems intended for real-world, high-stakes applications [6,9,12,16]. Empirical studies routinely involve older adults, caregivers and care staff through interviews, co-creation workshops, Wizard-of-Oz trials, and iterative in situ evaluations. After deployment, UCD supports continuous improvement of dialogue policies, escalation protocols, and personalisation. Although many studies still report usability and acceptability rather than clinical outcomes [10,18], co-designed systems tend to achieve higher engagement and smoother adoption. As a result, co-design is widely regarded as standard practice in this domain. We have likewise adopted this approach to gather requirements and iteratively evolve a simulator-based embodied care robot through successive workshops.

2.2 Existing Work on Conversational AI in Elder Care

Recent reviews show that conversational AI for older adults is used mainly in three ways: to monitor health and daily needs, to coach healthy behaviours, and to provide social companionship. Adoption often depends on trust, privacy protection, and whether users see clear benefits [1,2]. Case studies with the humanoid robot Nadine in nursing homes demonstrate short, engaging interactions with cognitively impaired residents, while also explores robustness challenges in complex care environments [14]. Reviews focused on socially assistive robots for people living with dementia find good feasibility and acceptability overall, but mixed or under-powered evidence for consistent clinical improvements, pointing to the need for longer, controlled studies in real homes and long-term care settings [7]. Complementing this literature, here we also examine the use of counterfactual explanations from XAI [3,8,15] as a basis for clarification dialogue, where a conversational agent could specify the minimal, user-verifiable changes under which a proposed action becomes safe (e.g., drying a wet floor before following through a command). This novel means of framing, operationalises hazard mitigation and dynamic consent, helping to balance safety with independence and, in turn, strengthening trust and uptake. We pursue this thread in our workshops, to understand when and how to incorporate counterfactual constructs into the robot's dialogue for clarification and hazard mitigation.

2.3 Simulators for Embodied Agent Research

To address the complex interactional requirements identified through co-design and overcome the limitations of monolithic systems, development and testing rely heavily on advanced simulation environments. Simulators like AI2-THOR [5] and VirtualHome [11] offer rich, interactive platforms for training and evaluating embodied agents (Fig. 1). AI2-THOR provides photorealistic indoor scenes for tasks like navigation and object manipulation, while VirtualHome focuses on modeling complex, multi-step household activities from natural language.

For this project, we selected the LEGENT simulator [4]. We selected LEGENT because its native chat interface allows users to issue natural language commands to an embodied agent. This feature is not only crucial for testing a conversational framework but also makes the system highly accessible for demonstrations with non-technical participants, such as the older adults and caregivers in our workshops. Furthermore, the modularity of LEGENT's chat pipeline allowed us to easily replace its default implementation with our custom multi-agent framework. This combination of an intuitive user interface and technical flexibility made LEGENT the ideal platform for developing, testing, and iteratively refining our system in alignment with the feedback gathered from our co-creation process.

Fig. 1. Simulators explored in this research: VirtualHome [11], AI2THOR [5] and LEGENT [4].

3 LLM Driven Robot Simulation

3.1 LEGENT: Architecture and LLM Integration

CoRDial's embodied interactions are grounded within LEGENT [4], an open platform for developing LLM-driven embodied agents in interactive 3D environments. LEGENT is structured around three components: a simulated environment, an agent model, and a data generation pipeline.

The **environment** provides real-time physics simulation (gravity, friction, collision dynamics), diverse rendering styles, and fully interactable objects including furniture, doors, and household items that agents can pick up, transport, place, and hand over. Scenes can be generated procedurally at scale or guided by natural language queries using LLMs, producing diverse indoor layouts without manual authoring.

The **agent** perceives the environment through a camera mounted at head height, communicates in natural language, and acts through a small set of continuous primitive actions: `move`, `rotate`, `interact`, and `speak`.

3.2 Adapting LEGENT for Personalised Elder Care

LEGENT's default configuration assumes a stateless interaction model: the LLM receives a scene description and a task, and generates a symbolic script. This is sufficient for general-purpose embodied tasks but does not address the requirements of complex dialogue necessary for personalised conversation. Specifically, the default system provides no mechanism for:

- Adapting responses to individual communication preferences (e.g., simplified language, preferred forms of address, or modality preferences)
- Maintaining continuity across sessions through user-specific episodic memory
- Disambiguating underspecified care requests (e.g., *"my tablets"* when multiple medications are scheduled)
- Applying care-specific safety constraints around medication, meals, and escalation to human carers
- Routing requests to appropriate specialist agents depending on care domain

Our primary contribution at the simulation layer was to address these gaps through **dynamic prompt construction**. Rather than providing the LLM with a fixed scene description and task string, CoRDial assembles the prompt at runtime by injecting several layers of context: the user's personal profile (communication preferences, care history, known preferences), the relevant care-domain constraints for the current request (e.g., scheduled medication, dietary restrictions, appointment state), and where necessary, a structured disambiguation turn that requests clarification before committing to an action script. This means the same LEGENT symbolic planning and motion execution pipeline operates unchanged underneath, while the behaviour of the LLM planner is shaped by user- and task-specific context at each interaction.

For example, a request such as *"Can you bring me my tablets?"* would in a default LEGENT session produce a generic medication retrieval script. In CoRDial, the prompt injected to the LLM additionally contains the user's current medication schedule, any safety flags (e.g., whether the evening dose has already been dispensed), and the user's preferred level of verbal explanation, producing behaviour that is both safe and individually tailored. If the request is ambiguous for instance when multiple items in the simulated world match "tablets" the constructed prompt instructs the LLM to generate a `speak()` clarification turn rather than proceeding to a `goto` command, deferring action until the user confirms their intent.

This separation of concerns, with LEGENT providing embodied grounding and CoRDial providing personalised context, meant that we could adapt the simulator's behaviour without modifying its core architecture. The design of the information injected into the LLM prompts was informed by a series of co-design workshops with older adults and care workers.

4 Co-creation Activity

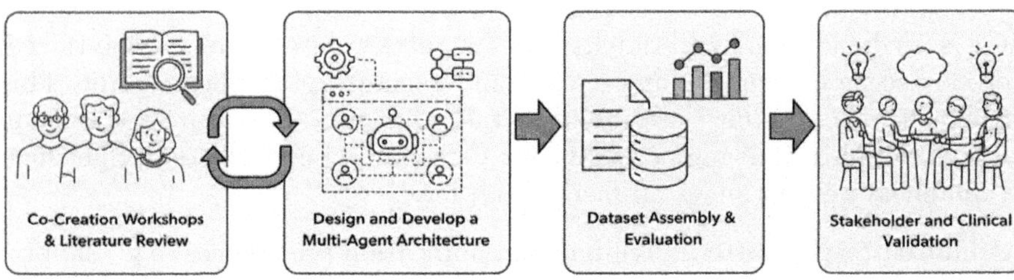

Fig. 2. Workflow from user-centered co-design to embodied prototype evaluation.

Figure 2 summarises the overall development workflow, showing how insights from user-centred co-design and literature review informed the design of the multi-agent architecture, the assembly of the evaluation dataset, and stakeholder and clinical validation. Seven workshops were conducted with 27 participants, including older adults aged 65 and above, informal caregivers, and professional caregivers. The study received ethical approval from the Research Ethics Committee of the School of Health Sciences at Robert Gordon University. Participants were recruited from a range of care contexts, including urban, rural, remote, and respite care settings. The workshops were informed by Patient and Public Involvement (PPI) and Participatory Action Research (PAR) approaches, ensuring that participants were actively involved in shaping the development of the system. Each workshop began with a brief introduction to the evolving prototype, followed by stimulus materials such as videos or live simulations demonstrating robot-assisted care tasks. Each session concluded with a guided group discussion in which participants reflected on the system's perceived usefulness, safety, communication style, and opportunities for personalisation.

Through these guided discussions, participants highlighted three key needs: proactive hazard management, natural and fluid conversation, and personalisation that remains under the users' control. They also wanted the robot to clearly explain its safety boundaries. Rather than simply refusing or failing to carry out a command, the robot should use counterfactual reasoning to explain why an action is unsafe, what would need to change, and what safer alternatives may be possible. These findings guided the decomposition of system functionality into specialised agents and informed the configuration of the overall agentic architecture.

5 Personalised Agentic Interactions

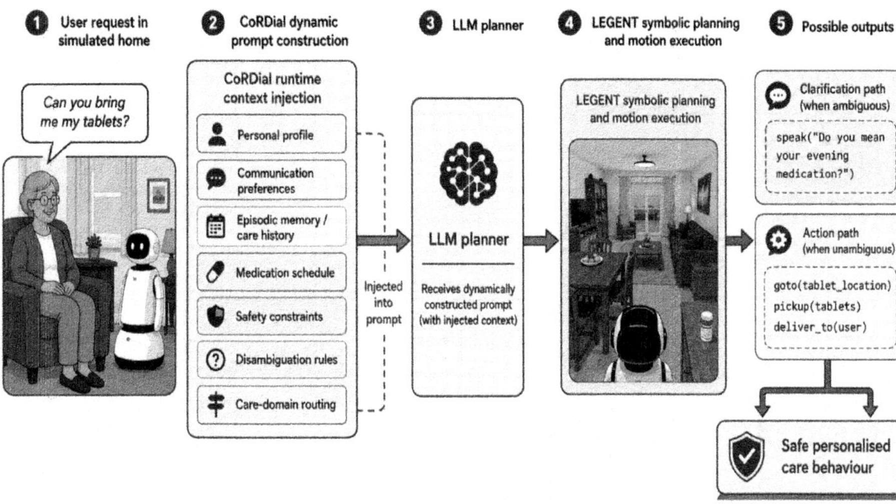

Fig. 3. CoRDial architecture for safe personalised care planning in a simulated home environment.

As shown in Fig. 3, the CoRDial architecture integrates dynamic prompt construction, LLM-based planning, and LEGENT symbolic execution to support safe and personalised care behaviour in a simulated home environment. Figure 4 presents the overall orchestration of the implementation pipeline. The process begins with the Planning Component, which uses a Conversational Intent Splitting LLM together with a ProtoNet Router to break down user utterances and classify them into actionable tasks. These tasks are then passed to the Execution Component, where specialised agents, including Simulation, Medication, Chit-chat, and Reminder agents, handle domain-specific requests.

The generated outputs are subsequently reviewed by the Ethics and Compliance Component, where an LLM-based reviewer evaluates candidate responses against relevant safety and care guidelines. Before the final response is delivered, the Personalisation Component adapts the wording and presentation to the users persona and accessibility profile which are saved in a database.

This workflow separates planning, execution, ethical review, and personalisation, enabling each stage to be inspected and modified independently. It also reflects the iterative refinements identified through the user co-creation activity, where improvements in disambiguation, multi-intent handling, and hazard detection emerged as primary areas to focus. Among these, multi intent handling directly motivated the development of an intent router, since user requests in elder care settings may combine several needs within a single utterance and must therefore be decomposed and directed to the appropriate downstream agents.

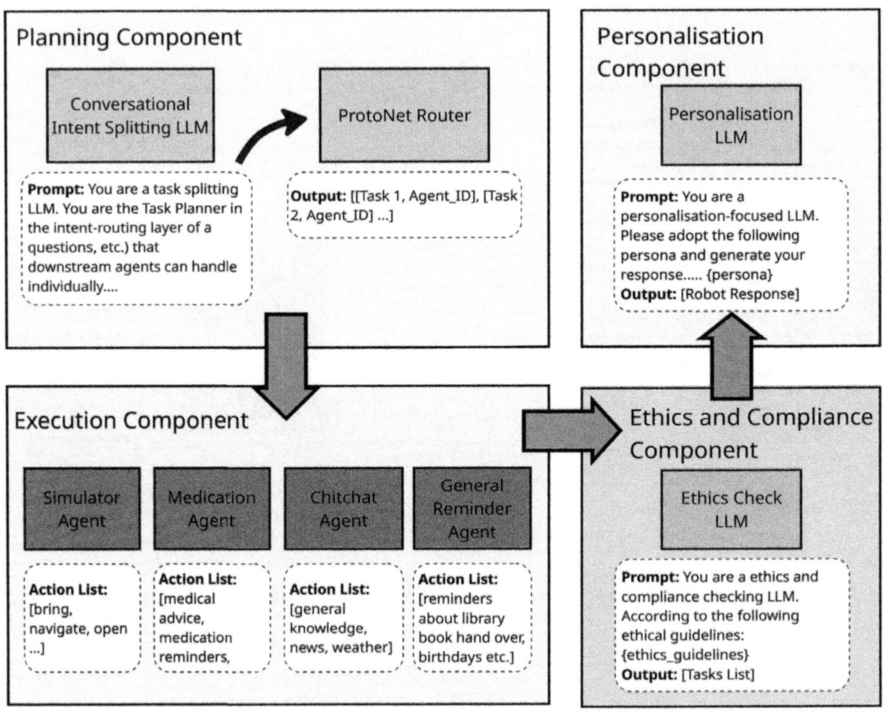

Fig. 4. Prompt flow across the planning, execution, ethics and compliance, and personalisation components.

6 Intent Router

In elder care settings, accurate intent recognition alone is not sufficient; routing decisions must also be fast, reliable, and explainable. Since users may phrase requests in varied or unexpected ways, and because delays can affect the usability of interactive support systems, domains such as eldercare require approaches that balances accuracy with low latency and transparent decision-making.

To address this, we use a class prototype based routing mechanism, referred to as ProtoKNN Router, which classifies user utterances using a few shot learning strategy. For each intent class c, a small labelled support set $S_c = \{x_1, \ldots, x_n\}$ is encoded using a sentence embedding model $f(\cdot)$. The class prototype p_c is then computed as the mean of the embedded support examples:

$$p_c = \frac{1}{|S_c|} \sum_{x_i \in S_c} f(x_i)$$

At inference time, an incoming utterance q is embedded using the same encoder and assigned to the class with the nearest prototype:

$$\hat{c} = \arg\min_{c} d(f(q), p_c)$$

where $d(\cdot)$ denotes a metric distance function, such as Euclidean distance. This formulation enables ProtoKNN router to avoid rigid keyword matching while retaining low latency and transparency, since routing decisions can be explained by the closest prototype and its associated support examples.

The current experiments evaluate ProtoKNN router under a single-intent classification setting, in which each utterance is assigned to its nearest intent prototype. This choice reflects the structure of the evaluation dataset, where each example is annotated with one primary intent, and allows the routing accuracy and latency of the prototype-based classifier to be assessed independently. In the full CoRDial architecture, however, multi-intent requests are handled before classification through the content-splitting stage shown in Fig. 4. The Conversational Intent Splitting LLM decomposes a compound utterance into smaller intent specific segments, such as separating "Can you bring my tablets and remind me to call my daughter later?" into a medication related request and a reminder related request. Each segment is then embedded and routed independently by ProtoKNN router to the appropriate specialist agent. This design preserves the simplicity and interpretability of nearest-prototype routing, since each routed segment remains explainable through its closest class prototype and associated support examples, while enabling multi-intent dialogue through decomposition and downstream orchestration. Thus, the single-intent evaluation isolates the performance of the ProtoKNN routing step, whereas multi-label routing in CoRDial is supported at the architectural level through content splitting followed by agent-level coordination.

The responsibility for deciding how the activated intents are handled is then separated from intent recognition itself. ProtoKNN router identifies the relevant intent classes, while the planning layer determines task ordering, agent assignment, and execution dependencies. Some intents may be handled independently by different agents and executed in parallel, whereas others may require sequential processing due to safety, contextual, or conversational dependencies. For example, a medication request may need to be resolved before a reminder is scheduled if the reminder depends on the medication identity or time. Conversely, a weather query and a social reminder can be handled in parallel. Thus, multi intent routing is treated in this paper as an architectural capability and methodological extension motivated by the co creation activity.

7 Evaluation

We tested the ProtoKNN router against several baseline families, each representing a different routing strategy:

- LLM based few shot routers, which use prompted generative models for direct intent prediction.
- Agent description baselines, which perform zero shot matching between utterances and agent descriptions.
- ProtoNets [13], which provide the closest prototype based baseline by learning the embedding function through episodic few shot training before computing class prototypes.
- Weighted ProtoNet variants [17], which adjust the contribution of support examples during prototype construction.
- ProtoKNN variants, which keep the sentence encoder fixed and perform direct nearest prototype matching using predefined distance metrics, as described in Sect. 6.

To evaluate the single intent routing mechanism, we created a synthetic but human-validated corpus specifically focused on elder-care interactions[1]. The dataset consists of 150 utterances, evenly distributed across three primary agent categories: Chit-chat, Medication Reminder, and Simulator commands. Semantic analysis of the dataset confirms that the structural diversity aligns with each agent's function. For instance, Chit-chat queries exhibit the broadest semantic range (lowest Average Pairwise Similarity (APS), of 0.1790), whereas the Medication Reminder queries are the most structurally focused (highest APS of 0.3418). This ensures our evaluation reflects both the open-ended nature of social interaction and the strict parameters of clinical requests.

Intent classification results for five families of intent classifiers are presented in Table 1. The LLM based few shot routers achieved the highest overall accuracy, with GPT 4o Mini and Gemini 2.5 Flash reaching up to 99.33%. However, the ProtoKNN variants showed strong improvement as the number of support examples increased. For example, ProtoKNN with Euclidean distance improved from 70.00% in the 1 shot setting to 94.00% in the 10 shot setting, while the Mahalanobis variant increased from 70.67% to 96.00%. This substantially outperformed the agent description baselines, which achieved only 33.33% – 44.67% accuracy in the zero shot setting. Compared with ProtoNets and weighted ProtoNet variants, ProtoKNN achieved competitive but slightly lower accuracy in most settings, with its strongest result obtained by the Mahalanobis variant at 10 shots. Although ProtoKNN does not achieve the highest accuracy overall, it offers a practical deployment advantage over trained prototype based models such as ProtoNets. New intent classes can be incorporated by encoding their labelled support examples and recomputing the corresponding class prototype, without retraining the embedding model. This makes ProtoKNN particularly

[1] https://doi.org/10.5281/zenodo.20650275.

suitable for extensible routing scenarios where new elder care intents or agents may be added over time.

It is worth noting that ProtoKNN router's accuracy is highly sensitive to the number of support examples. At 1-shot, all ProtoKNN variants achieve only 70.00–70.67%, which is notably lower than LLM-based routers in the same setting. This is likely because a single support example provides limited coverage of the linguistic variation within an intent class. As the resulting prototype is static, it may not capture the broader range of semantically related utterances that should map to that class. Performance improves substantially as shot count increases, with the Mahalanobis variant reaching 96.00% at 10-shot. This reliance on support set size has practical implications: in deployment contexts where only a small number of labelled examples are available for each intent class, ProtoKNN may need to be supported by additional data collection before it can be used reliably for routing.

Table 1. A comparison of various few-shot learning models across different shot counts.

Approach/Model	0	1	3	5	10
Few-Shot (GPT-4o-Mini)	98.67	99.33	99.33	99.33	98.67
Few-Shot (Gemini 2.5 Flash)	97.33	98.67	99.33	99.33	99.33
Few-Shot (Llama3 8B Instruct)	98.67	97.33	99.33	98.00	98.00
AgentDescription (Euclidean)	44.67	N/A	N/A	N/A	N/A
AgentDescription (Cosine)	43.33	N/A	N/A	N/A	N/A
AgentDescription (Mahalanobis)	33.33	N/A	N/A	N/A	N/A
ProtoNets	N/A	N/A	88.67	95.33	94.67
ProtoKNN (Euclidean)	N/A	70.00	80.00	92.67	94.00
ProtoKNN (Cosine)	N/A	70.67	80.00	92.67	94.00
ProtoKNN (Mahalanobis)	N/A	70.67	87.33	93.33	96.00
Weighted ProtoNets (BMP)	N/A	N/A	88.67	94.00	95.33
Weighted ProtoNets (ANPC)	N/A	N/A	88.00	92.67	95.33

Table 2. Per-class F1 comparison between LLM-based routers and ProtoKNN in the 10-shot setting

Router	Chit-chat	Medication	Simulator	Macro-F1
GPT-4o-mini	1.000	0.980	0.980	0.987
Gemini 2.5 Flash	1.000	0.990	0.990	0.993
Llama3-8B	1.000	0.970	0.970	0.980
ProtoKNN (Mahalanobis)	0.989	0.950	0.940	0.960

Table 3. End-to-End Model Latency Comparison (in milliseconds). Latency was measured at the 5-shot setting as a representative configuration

Model	Mean Latency (ms)	Median Latency (ms)	Std. Dev. (ms)	Min Latency (ms)	Max Latency (ms)
Few-Shot (GPT-4o-Mini)	460.31	419.42	118.01	317.84	773.22
Few-Shot (Gemini 2.5 Flash)	470.14	442.97	165.97	269.85	1002.70
Few-Shot (Llama3 8B Instruct)	380.24	376.41	34.55	325.08	472.62
AgentDesc. (Euclidean)	106.40	105.76	12.28	88.97	136.42
AgentDesc. (Cosine)	106.32	105.69	12.28	88.88	136.34
AgentDesc. (Mahalanobis)	108.36	107.58	12.28	90.84	138.21
Protonets	133.39	133.14	12.93	114.81	170.64
ProtoKNN (Euclidean)	106.35	105.72	12.28	88.90	136.36
ProtoKNN (Cosine)	106.30	105.67	12.28	88.85	136.31
ProtoKNN (Mahalanobis)	108.20	107.60	12.29	90.69	138.16
Weighted Protonets (BMP)	133.47	133.10	12.27	117.05	163.67
Weighted Protonets (ANPC)	133.07	132.27	13.16	114.74	164.10

Although ProtoKNN does not achieve the highest accuracy overall, it offers a practical deployment advantage over trained prototype based models such as ProtoNets. New intent classes can be incorporated by encoding their labelled support examples and recomputing the corresponding class prototype, without retraining the embedding model. This makes ProtoKNN particularly suitable for extensible routing scenarios where new elder care intents or agents may be added over time.

Because different intent classes have different safety implications in care settings, we additionally report per-class F1 scores for the 10-shot setting (Table 2). The LLM-based routers achieve the strongest class-wise performance, with Gemini 2.5 Flash obtaining the highest macro-F1 of 0.993. ProtoKNN (Mahalanobis) achieves a lower but still competitive macro-F1 of 0.960, with F1 scores of 0.989 for Chit-chat, 0.950 for Medication Reminder, and 0.940 for Simulator commands. Importantly, the Medication Reminder class, which is the most safety critical category in this evaluation, remains strong under ProtoKNN, although LLM-based routers achieve higher recall and F1.

The comparison with LLM based routers further highlights this trade off. The LLM routers achieve the strongest accuracy results in Table 1, but this improvement comes at a substantial latency cost. Crucially, by computing metric distances to prototype embeddings rather than relying on generative LLM inference for routing, ProtoKNN substantially reduces system latency, as shown in Table 3. While the LLM routers achieved higher accuracy, they incurred approximately 3.6 to 4.4 times higher mean latency than ProtoKNN. This suggests that

ProtoKNN provides a more suitable trade off for latency sensitive deployment in the elder care domain.

8 Conclusion

CoRDial demonstrates how a modular multi agent architecture can support safer and more personalised conversational interaction in elder care robotics without depending on a single monolithic LLM. By separating dialogue, planning, ethical review, and simulated physical execution, the framework provides a flexible basis for coordinating care related tasks while maintaining explicit points for safety checking and contextual adaptation. The evaluation of ProtoKNN router further shows that efficient intent routing is central to this architecture. Although LLM based routers achieved the highest accuracy, ProtoKNN router offered a stronger latency accuracy trade off by using prototype based distance matching rather than generative inference. This makes it a practical routing mechanism for latency sensitive elder care scenarios, particularly where new intents or agents may need to be incorporated over time. The sensitivity of ProtoKNN to shot count also highlights a practical open challenge: in real care deployments, obtaining sufficient labelled examples for every intent may not always be feasible. Future work should explore dynamic prototyping, active and continual learning, and synthetic data augmentation to improve low shot performance, while evaluating routing robustness on more naturalistic, user generated utterances and monitoring for possible routing drift.

Acknowledgement. This research is supported by the EMERGENCE EPSRC Healthcare Technologies Network+ through the Robotics for Frailty Challenge grant.

GenAI Usage Disclosure. Initial layout schematics for Fig. 3 were drafted using the AI-assisted tool Gemini. These layouts were subsequently manually created in Canva and detailed components added and verified by the authors to ensure accuracy.

References

1. An, J., Yi, S., Lyu, Y., Liu, H., Zhang, Y.: Conversational agents for older adults' health: a systematic literature review. arXiv preprint arXiv:2503.23153 (2025)
2. Brownson-Smith, R., et al.: Conversational systems for social care in older adults: protocol for a scoping review. JMIR Res. Protocols **14**(1), e72310 (2025)
3. Byrne, R.M.: Counterfactuals in explainable artificial intelligence (XAI): evidence from human reasoning. In: IJCAI. pp, 6276–6282 (2019)
4. Cheng, Z., et al.: LEGENT: open platform for embodied agents. In: Cao, Y., Feng, Y., Xiong, D. (eds.) Proceedings of the 62nd Annual Meeting of the Association for Computational Linguistics (Volume 3: System Demonstrations), pp. 335–345. Association for Computational Linguistics, Bangkok, Thailand (2024). https://doi.org/10.18653/v1/2024.acl-demos.32, https://aclanthology.org/2024.acl-demos.32/
5. Deitke, M., et al.: RoboTHOR: an open simulation-to-real embodied AI platform. In: CVPR (2020)

6. Ghosh, R., et al.: User-centered design of socially assistive robotic combined with non-immersive virtual reality-based dyadic activities for older adults residing in long term care facilities. arXiv preprint arXiv:2410.21197 (2024)
7. Hsieh, C.J., Li, P.S., Wang, C.H., Lin, S.L., Hsu, T.C., Tsai, C.M.T.: Socially assistive robots for people living with dementia in long-term facilities: a systematic review and meta-analysis of randomized controlled trials. Gerontology **69**(8), 1027–1042 (2023)
8. Keane, M.T., Smyth, B.: Good counterfactuals and where to find them: a case-based technique for generating counterfactuals for explainable AI (XAI). In: Watson, I., Weber, R. (eds.) ICCBR 2020. LNCS (LNAI), vol. 12311, pp. 163–178. Springer, Cham (2020). https://doi.org/10.1007/978-3-030-58342-2_11
9. Maxwell, C., et al.: User-centered design for socially assistive robotic activities with older adults in long term care. Innov. Aging **7**(Suppl 1), 1149 (2023)
10. Milne-Ives, M., et al.: The effectiveness of artificial intelligence conversational agents in health care: systematic review. J. Med. Internet Res. **22**(10), e20346 (2020)
11. Puig, X., et al.: VirtualHome: simulating household activities via programs. In: Proceedings of the IEEE Conference on Computer Vision and Pattern Recognition, pp. 8494–8502 (2018)
12. Salomé, S., Du Bousquet, L., Monfort, E.: Co-designing conversational agents with older people: a scoping review of methods, challenges, and a way forward. Comput. Hum. Behav. Rep. **17**, 100606 (2025)
13. Snell, J., Swersky, K., Zemel, R.: Prototypical networks for few-shot learning. Advances in Neural Information Processing Systems, p. 30 (2017)
14. Tulsulkar, G., Mishra, N., Thalmann, N.M., Lim, H.E., Lee, M.P., Cheng, S.K.: Can a humanoid social robot stimulate the interactivity of cognitively impaired elderly? A thorough study based on computer vision methods. Vis. Comput. **37**(12), 3019–3038 (2021)
15. Wachter, S., Mittelstadt, B., Russell, C.: Counterfactual explanations without opening the black box: automated decisions and the GDPR. Harv. JL Tech. **31**, 841 (2017)
16. Wegener, E.K., M Bergschöld, J., Kramer, T., Schmidt, C.W., Borgnakke, K.: Co-designing a conversational agent with older adults with chronic obstructive pulmonary disease who age in place: qualitative study. JMIR Hum. Factors **11**, e63222 (2024)
17. Wijayasekara, V.A., Martin, K., Wiratunga, N., Massie, S., Wijekoon, A.: Few-shot essay grading: weighted prototypical networks for ordinal text classification. In: Lynce, I., et al. (eds.) ECAI 2025: Proceedings of the 28th European Conference on Artificial Intelligence 2025, Including the 14th Conference on Prestigious Applications of Intelligent Systems. Frontiers in Artificial Intelligence and Applications, vol. 413, pp. 2155–2162. IOS Press, Amsterdam (2025). https://doi.org/10.3233/FAIA251055
18. Yu, C., Sommerlad, A., Sakure, L., Livingston, G.: Socially assistive robots for people with dementia: systematic review and meta-analysis of feasibility, acceptability and the effect on cognition, neuropsychiatric symptoms and quality of life. Ageing Res. Rev. **78**, 101633 (2022)

Robotic Pet Companions Integrated with Large Language Models for Cognitive, Emotional, and Social Engagement of Elderly Persons

Khaja Mohteshamuddin[1(✉)], Ayesha Ahmad[1], and Ali Nawaz[2]

[1] College of Agriculture and Veterinary Medicine, United Arab Emirates University, P.O. Box 15551, Al Ain, United Arab Emirates
{drkhaja707,700051104}@uaeu.ac.ae

[2] College of Information Technology, United Arab Emirates University, P.O. Box 15551, Al Ain, United Arab Emirates
700040004@uaeu.ac.ae

Abstract. The rapid growth of the elderly population worldwide has worsened problems of social isolation, cognitive decline, and emotional health. Existing LLM-based conversational AI offers some support for companionship and monitoring, but it is screen-bound and cannot reproduce the physical and emotional benefits of human-pet contact. This paper proposes a framework that combines LLM-driven conversation with robotic pets to support cognitive, emotional, and social engagement in elderly users. Two deployment modes are introduced: a single-user mode, in which one robotic pet provides personal companionship and continuous monitoring, and a group mode, in which animal-like companions act as social catalysts in shared care settings. The system includes a Dialogue Manager, Memory Module, Cognitive and Emotional Testing Engine, Robotic Pet Interface, Analytics Engine, and a Caregiver Dashboard. With user consent, family social media content is relayed through the pet to maintain family connectedness. Compared with digital-only LLM agents and humanoid robots, the proposed approach offers higher user acceptance, avoids the uncanny valley, supports passive monitoring via touch and proximity sensors, and provides continuous animal-assisted-style therapy. Mixed-methods evaluation of cognitive scores, sentiment trend analysis, logs of robotic engagement, and caregiver feedback is also proposed.

Keywords: Animal assisted therapy · Social wellbeing · Psychosocial rehabilitation · Human-animal bond · LLM

1 Introduction

The global population of elderly individuals is growing rapidly. By 2050, it is estimated that there will be over 1.5 billion people aged 65 years and above, posing serious challenges for healthcare and social services [33]. Among the main

S. S. Khan et al. (Eds.): IJCAI 2026, CCIS 3127, pp. 91–106, 2027.
https://doi.org/10.1007/978-981-92-5190-2_7

issues faced by individuals of this age group are social isolation, which impacts nearly one-third of elderly adults in developed nations, and cognitive impairment associated with mild cognitive impairment (MCI) and dementia [16]. Both of these concerns interact with one another and contribute to cognitive and emotional issues. In particular, prolonged loneliness increases the likelihood of cognitive deterioration, depression, cardiovascular disease, and premature mortality [2]. Consequently, elderly individuals require assistance that is cognitively stimulating, emotionally comforting, and socially engaging.

Current solutions are inadequate. Clinical assessment occurs rarely, is expensive, and can induce stress in patients, undermining its reliability [13]. While rule-based chatbots can be scaled easily, they lack the capacity for deeper and more memorable engagement [18]. On the other hand, brain training games such as Lumosity provide cognitive stimulation but fail to address emotional and social factors [27]. Humanoid robots provide a physical presence, yet many of them elicit the effect known as the Uncanny Valley, where almost human-like robots are perceived to be uncanny [19].

With recent advances in Large Language Models (LLMs) such as GPT-4 and LLaMA 2, a novel type of companion can now be created. Such companions are capable of engaging in rich, empathic conversation, integrating cognitive testing in dialogue, and developing a long-term profile of the user [9,24]. However, despite these advantages, they are still constrained by the screen barrier. They lack the ability to provide a touch, physical presence, and comforting presence of a companion.

Research on animal-assisted therapy suggests that interaction with both real and robotic animals is associated with reduced levels of cortisol, decreased anxiety, improved mood, and increased social engagement [4]. In turn, robotic pets such as PARO (a robotic seal) and Sony's AIBO (a robotic dog) have been used successfully to help older adults, including those diagnosed with dementia, to overcome loneliness without succumbing to the Uncanny Valley due to animal form [26,35]. However, while these pets are helpful, they are not able to engage in actual conversation, remember past encounters, or adjust to changing user condition.

In this paper, a framework that integrates these two strands of research is proposed. In particular, we propose a solution that leverages LLMs to enable rich and empathic conversations with older adults while integrating the advantages of robotic pets as physical entities. Two deployment modes are defined: single-user pet companionship for personal attachment and monitoring and group robotic pet facilitation that utilizes the pet as a social catalyst in group settings. The system also helps to keep users updated on their families through family-related posts on social media relayed through the pet.

Contributions of this paper include the following: (i) framework for integrating LLMs and robotic pet companions for elderly care; (ii) dual deployment mode design for both single-user and group utilization; (iii) comprehensive comparative analysis of digital companions, humanoid companions, and our system; (iv) cognitive and emotional monitoring pipeline based on dialogue and pet sensor data; and (v) mixed-methods evaluation approach.

The rest of this paper is organized as follows. Section 2 reviews related work. Section 3 presents the architecture and deployment modes. Section 4 describes cognitive and emotional monitoring. Section 5 covers privacy and ethics. Section 6 outlines the evaluation plan. Section 7 discusses limitations, and Sect. 8 concludes.

2 Related Work

2.1 Conversational AI and LLMs in Elderly Care

Conversational AI in elderly care has shifted from rigid rule-based systems [18] to transformer-based LLMs. Models such as GPT-4 [24] and LLaMA 2 [32] support nuanced, context-aware dialogue that can be sustained across sessions through memory. Xygkou et al. [37] developed MindTalker for early-stage dementia, while Raile [22] reported on the use of ChatGPT in therapy. Earlier work such as Ellie at USC ICT showed that virtual agents can conduct structured clinical interviews for PTSD [21]. Replika has also shown that users can form meaningful bonds with conversational AI [8]. However, none of these systems are tailored to elderly users, include physical embodiment, or combine dialogue with clinical monitoring in a unified way.

2.2 Humanoid Robots and the Uncanny Valley

Humanoid robots such as NAO and Pepper have been used in eldercare for exercise, memory games, and social interaction [1,23,31]. Yet they often trigger the Uncanny Valley described by Mori [19], in which near-human but imperfect forms cause unease, particularly for users with cognitive impairment [7]. They are also costly and complex, with rigid scripted dialogue. Interestingly, Sung et al. [29] found that even the non-anthropomorphic Roomba sparked social conversation in households, suggesting that strong social facilitation does not require a human-like form and may actually be helped by a non-human design.

2.3 Robotic Pets and Animal-Assisted Therapy

The mental-health benefits of animal-assisted therapy (AAT) for older adults are well established. Beetz et al. [4] showed that interacting with animals reduces cortisol and loneliness while encouraging social activity. Volunteer pet visitation has similar effects in care facilities [5], but is constrained by hygiene rules, allergies, animal welfare, and scheduling. Robotic pets address many of these issues. PARO, the most studied robotic pet, has been shown to reduce stress and improve mood among care-home residents and to spark conversation between people who otherwise rarely interacted [35]. AIBO has helped residents with severe dementia engage more actively compared with a stuffed toy or no intervention [30]. Sharkey and Sharkey [25] raise important ethical concerns about deception and dignity, which we take seriously through transparency and consent. The shared limitation of existing robotic pets is that they rely on scripted reflexes, not real conversation. Adding LLM intelligence is the natural next step.

2.4 Social Media and Family Connectedness

Cotten et al. [11] linked internet use among older adults to a 30% drop in depression risk, mostly through better social connectedness. Family group chats on WhatsApp or Facebook are now central to family life, yet many elderly users are passive observers or absent altogether, which deepens their sense of exclusion [10]. According to Vines et al. [34], older individuals desire technology that would enable them to remain connected to their families, rather than simply technology itself. Our framework enables this through the use of the robotic pet as an intermediary that shares relevant news from their family members with their permission.

2.5 Positioning of This Work

To the best of our knowledge, there is currently no existing work that incorporates (i) LLM intelligence into robotic pet form, (ii) the designations of single-user and group deployment models, (iii) the use of family social media to overcome digital exclusion, and (iv) a clinical assessment framework to measure cognitive, emotional, and social effects. Table 2 (Sect. 3.5) compares the proposed solution to its nearest competitors.

3 System Architecture

Our framework is designed as a modular, microservices-oriented architecture that facilitates real-time personalized interactions, both one-on-one and group-wise. The key innovation within our framework lies in the concept of Robotic Pet Interface Layer that connects the physical pet to the LLM core and the monitoring pipeline. Figure 1 illustrates the architecture; the components are described below.

3.1 Core System Modules

User Interface. The interface runs on smartphones, tablets, and smart home devices, with both voice and text input. Accessibility features include large fonts, high contrast, voice navigation, and a simplified companion app for reviewing past sessions and memory notes. Flutter and React Native are used for cross-platform support on iOS and Android.

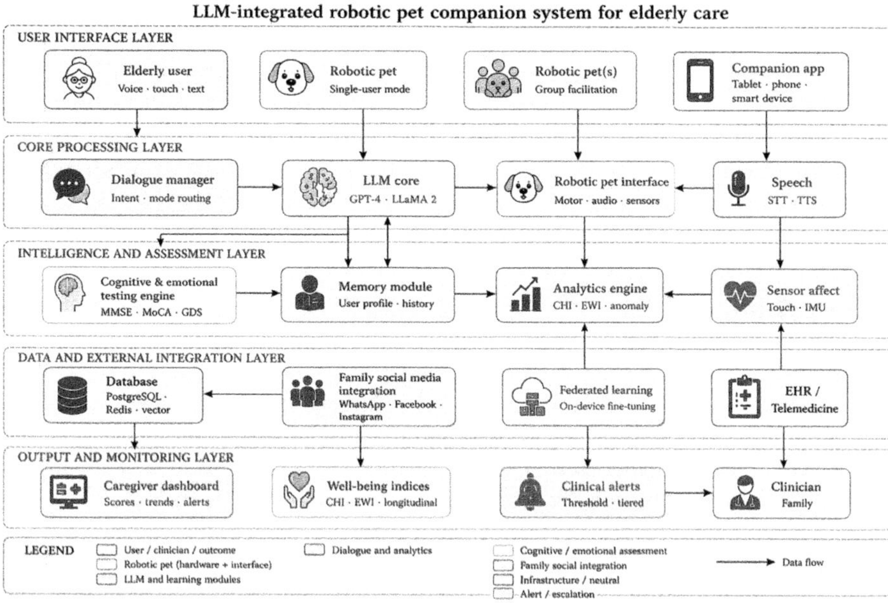

Fig. 1. LLM integrated robotic pet companion system architecture for elderly care.

Robotic Pet Interface. This is the core innovation of the framework. It works as a bidirectional middleware layer between the LLM Core and supported pet hardware (Sony AIBO, AIST PARO, and open-platform companions) using ROS 2 and MQTT. Outbound, it turns LLM output and affective intent into motor commands, LED patterns, and synthesized speech routed through the pet's speaker. Inbound, it gathers touch, accelerometer, proximity, microphone, and optional camera data, and passes it to the LLM Core and Analytics Engine.

The system is hardware agnostic as the addition of any new platform is achieved via driver modules that use the same data schemas. In the absence of physical hardware, the pet may be rendered as an animated avatar on a tablet while still retaining all functionality associated with the dialogue and monitoring.

Dialogue Manager. The Dialogue Manager deals with recognizing user intent and dispatching the inputs in one of the four modes: (i) Companion Mode, for open-ended empathic dialogue; (ii) Assessment Mode, for embedding cognitive and emotional tests into the dialogue flow; (iii) Family Update Mode, for relaying family social media content through the pet; and (iv) Caregiver Relay Mode, for medication reminder and other caregiver-related information. A finite state machine along with intent classifier based on LLM ensures seamless switching between these modes.

LLM Core. The LLM Core uses a fine-tuned transformer model (GPT-4 or LLaMA 2) tailored for elderly-friendly dialogue via specialized training on geriatric dialogue datasets, clinical transcripts, and therapeutic dialogue patterns. Such adaptations include simplified vocabulary, slower pace, culturally appropriate references, identity reinforcement, and empathy mirroring without amplification of negative emotions. LLM Core outputs behavioral intents (warmth, attentiveness, concern, playful demeanor) to the Robotic Pet Interface, which translates them into robot behavior.

Memory Module. The Memory Module keeps track of user's profile with details such as their family, hometown, professional background, preferences, baseline test results, past conversations, and past interactions with the pet. The database includes PostgreSQL tables for structured data, Redis cache for context within the current conversation session, and vector store for semantic search. It enables LLM Core to mention previous topics in conversation: 'Last time we talked about your granddaughter coming to visit, how did that go?'

Analytics Engine. The Analytics Engine performs analysis on dialogue content, prosody, cognitive scores, and pet sensor data to generate longitudinal trends and real-time alerts. It tracks week-by-week changes in sentiment, dialogue fluency, response latency, cognitive scores, and frequency of interactions with the pet, comparing them against the user's own baseline. Furthermore, it generates Cognitive Health Index (CHI) and Emotional Well-being Index (EWI) scores for further clinical interpretation.

Cognitive and Emotional Testing Engine. This module integrates clinical testing instruments, such as MMSE, MoCA, and Geriatric Depression Scale (GDS), into the regular conversation flow. Questions are reformulated as friendlier prompts, adjusted for user's state, and dispersed across different sessions. Calculated scores are saved in the Memory Module. Behavior of the robotic pet, such as initiating interaction and petting duration, serves as additional indirect measure of cognitive and emotional health.

Family Social Media Integration. Upon obtaining explicit consent from both the user and their family members, the system connects to group channels on WhatsApp, Facebook, and Instagram via secure API. With the help of Named Entity Recognition (NER) and summarization pipeline, personally relevant updates (events, birthdays, shared photos, etc.) are extracted and anonymized. The pet then shares the news warmly: 'Today your son posted a picture of your grandchildren at the beach. Would you like me to tell you more about it?'

Caregiver Dashboard. The Caregiver Dashboard is a secure web-based dashboard intended for clinicians and family members. It displays longitudinal charts of cognitive and emotional scores, pet interaction data, summaries of past conversations, and configurable alerts. Family members can add notes and export patient data in HL7 FHIR format for incorporation into Electronic Health Records (EHR). Access control allows family members to see a simplified version of the dashboard while clinicians get full analytics view.

3.2 Robotic Pet Deployment: Single-User and Group Configurations

An important structural innovation introduced by this project is the explicit segmentation into two deployment modes depending on the care setting.

Single User Robotic Pet Companion. The single user companion operates under three different but smoothly transitioning modes. The first mode, active dialogue mode, involves an interactive dialogue with the user complete with assessments and updates on family members, along with physical responses, such as tail wagging, head nodding, and sound effects. Passive presence mode involves maintaining the pet's activity quietly with breathing sounds, slight movements, and proximity-based touch interactions.

The Memory Module stores the complete profile of the user while the LLM Core is tuned according to the user's cognitive abilities and preferred language. This mode helps during rest, low energy, or mild agitation. Caregiver relay mode delivers reminders and messages in a personalized, non-clinical tone.

This mode suits users at home, those with moderate to severe isolation, individuals in early-to-moderate cognitive decline, and users who tend to bond with personal companions. Prior work suggests that personal ownership of a companion is linked to stronger mental-health outcomes than shared institutional resources [26].

Group Robotic Pet Facilitation. In group mode, one or more robotic pets are placed in shared spaces in care facilities, day centers, or community halls. The design draws on the social catalyst hypothesis: a novel, responsive, animal-like presence in a shared space tends to draw people together and lower the barriers to conversation [29]. The LLM Core in this mode runs a multi-participant session manager that tracks individual users and coordinates a shared dialogue theme across active pet units.

Three group activities are supported. In Group memory circles, the pet hosts gentle prompts such as 'I once saw a dog chasing its tail in the park, did any of you ever have a dog?' and invites quieter participants to join in. In collaborative care tasks, participants take turns 'feeding' or attending to the pet through simple actions, building a sense of shared purpose. The social bridging mode involves the use of the pet as a neutral medium that helps to re-engage disengaged participants or to introduce new topics.

An enhanced version involves the provision of a pet to each participant and the coordination of a common theme by the LLM Core, whereby each pet tells its own piece of a common story to be discussed by the participants together. Such a combination is capable of coping with diverse cognitive and communication capabilities of real care group members.

3.3 Advantages of Robotic Pets in the Proposed Framework

Robotic pets offer the following advantages over purely digital companions and humanoid robots:

- Physical interactions such as petting and holding a robot pet with its soft body can induce physiological reactions, like lowering of the heart rate and reducing levels of cortisol [4]. In particular, it can help dementia sufferers whose capacity for verbal communication may have been lost but physical interaction retained.
- Robotic pets shaped as animal-like characters (dogs, cats, seals, birds, etc.) are well accepted by elderly patients with dementia and even when they realize the artificial nature of the pet [35].
- Unlike real-life pets, robotic animals do not require feeding and medical services, as well as being easy to accommodate in places where hygiene and allergen requirements prevail.
- Continuous data about the user's behavior obtained by the sensors of the pet complement dialogue assessment of the patient's mental state without any additional effort from the side of the user.
- Interaction with the pet helps depressed, withdrawn, or cognitively impaired patients to get engaged in interactions with the AI companion and others.
- The shape, names, and personalities of the robot pets may be selected to correspond to recognizable animals, which allows for increased bonding that cannot be achieved by any static interface.
- When using a robot pet in public space, the pet becomes an object of interaction between peers without the necessity to speak with each other, which is helpful for people with communication disorders.
- The animal volunteer programs require a certain schedule and trained volunteers. The robotic pets can serve as an alternative to people on weekends and nights.

3.4 Technical Stack

Table 1 lists the technical tools used in each component, with notes on single-user and group use.

Table 1. Comprehensive Technical Stack of Modules, Tools, and Deployment Context

Component	Tools/Platform	Description	Single-Pet Use	Group-Pet Use
LLM Core	GPT-4, LLaMA 2, Mistral	Fine-tuned transformer for empathetic dialogue, memory recall, and language analysis.	Personalized to user history.	Coordinated multi-pet dialogue.
Robotic Pet Interface	AIBO SDK, PARO API, ROS 2, MQTT	Translates LLM outputs to motor, LED, and audio commands; feeds sensor data back.	Active and passive presence modes.	Synchronized behavior across pets.
Speech Modules	Whisper STT, ElevenLabs, Google TTS	Real-time speech-to-text and expressive voice synthesis.	Via pet speaker.	Per-pet audio with shared themes.
Sensor & Affect	Capacitive touch, IMU, optional camera	Passive data on touch, proximity, posture for behavior change detection.	Personal interaction logs.	Group heatmaps; flags withdrawn users.
Dialogue Manager	Rasa, LangChain, custom FSM	Intent routing between chat, assessment, family updates, and reminders.	Caregiver Relay and Passive Presence.	Social Bridging, round-robin.
Memory Module	PostgreSQL, Redis, vector store	Long-term profile: family, hobbies, prior scores, pet preferences.	Pet recalls user stories.	Shared themes plus sub-profiles.
Analytics Engine	Python, scikit-learn, HuggingFace	Trend analysis and anomaly detection from speech and sensor data.	Single-user decline modeling.	Group-level engagement analytics.
Cognitive Testing Engine	MMSE/MoCA wrappers, NLP scorers	Embeds clinical instruments as natural prompts and auto-scores.	Adaptive pacing per user.	Group trivia with embedded scoring.
Family Social Integration	Facebook, WhatsApp, Instagram APIs	Extracts family updates and relays them through the pet.	Personalized messages.	Shared family news.
Caregiver Dashboard	React, FastAPI, MongoDB	Clinical portal for scores, alerts, and EHR annotation.	Individual timeline.	Group reports and heatmaps.
Frontend UI	Flutter, React Native	Accessible companion app with large fonts and simple navigation.	Personal session log.	Group activity board.
Backend & Data	FastAPI, Node.js, Docker	Microservices with encrypted storage and federated learning.	Edge inference for privacy.	Central multi-unit coordination.

3.5 Comparative Analysis

Table 2 compares the framework with two main alternatives: digital-only LLM agents and humanoid robots, across twelve dimensions relevant to elderly care.

Table 2 demonstrates that the proposed solution combines depth of conversation, physical presence, and high user acceptance without the drawbacks associated with Uncanny Valley in the case of humanoids. The distinctive features include the ability to facilitate groups, passively monitor sensors, and culturally adapt.

Table 2. Comparative Analysis of LLM + Robotic Pet vs. Digital-Only LLM vs. Humanoid Robot

Feature/Criterion	LLM + Robotic Pet (This Work)	Digital-Only LLM Agent	Humanoid Robot
Physical Tactile Presence	Yes-zoomorphic form	No	Yes-humanoid form
Uncanny Valley Risk	Low, due to animal form	None	High
Social Isolation Reduction	Individual and group modes	Individual only	Individual, limited group
Cognitive Assessment	In dialogue and pet behavior	In dialogue	Limited or scripted
Emotional Monitoring	Voice, language, and touch	Voice and language	Voice and vision
Group Facilitation	Yes-multi-pet coordination	No	Partial
AAT Substitute	Yes-continuous and hygienic	No	Partial
Cultural Adaptability	Pet form and LLM language	LLM language only	Appearance and language
Passive Behavioral Monitoring	Touch, proximity, interaction logs	Usage logs only	Vision-based
Privacy Preservation	Federated learning, edge inference	API-dependent	Embedded compute
Deployment Cost	Moderate (robot and cloud)	Low, software only	High
Acceptance (Elderly)	High, familiar animal form	Moderate	Variable

4 Cognitive and Emotional Monitoring

Clinical monitoring plays an essential part in the framework. Instead of conducting assessment at certain moments, continuous monitoring of the users' cognitive functioning and emotions occurs through regular interactions with the system. Data from the users is gathered via three channels: dialogue, voice prosody, and robotic pet sensor signals.

4.1 Embedded Cognitive Assessment

The Testing Engine chooses the tasks for the user from standardized tests like MMSE and MoCA. The tasks are presented to the user in natural language through dialogue. The tasks are spread evenly during the session to avoid fatigue, their difficulty and pace vary according to the user's condition, and the following types of assessments are made:

- Short-term memory: e.g., What did you eat for dinner yesterday?', to evaluate episodic memory informally.
- Executive function and verbal fluency: e.g., Name five things that you could find in a kitchen', to assess category fluency and narrative skill.
- Orientation: e.g., What season is it?', to detect orientation impairments, a typical sign of decline.
- Attention and processing speed: latency and turn-topic coherence are automatically analyzed.
- Pet indicators: Does the user start interaction independently? Can the user remember the pet's name? Does the user react to attempts of the pet to get its attention?

4.2 Emotional Assessment and Well-Being Monitoring

The emotional state of the user is assessed through the dialogue sentiment, voice prosody, behavioral pattern, and gentle mood probing:

- Dialogue sentiment: sentiment classifiers based on transformers are used to detect the prolonged negative sentiment and rapid change in it as a symptom of depression or anxiety [15].
- Voice prosody: pitch, rate, pause duration, and energy level are evaluated compared to the user's baseline. A flat prosody and a decreased energy level are indicative of low mood or fatigue [28].
- Repetitiveness: a sudden increase in topic repetition and the repetition of certain phrases is indicative of fixation or anxiety, or even dementia [20].
- Gentle mood probing: the LLM initiates light questions like You seem a bit quiet today, is everything ok?' after detecting negative sentiment.
- Pet interaction patterns: a decrease in touch duration or frequency and transition from active to passive interaction are indicative of withdrawal and are escalated to the Analytics Engine.
- Optional computer vision: where permitted, facial expression and posture analysis provide an additional data stream [14].

4.3 Integrated Well-Being Scoring

The Analytics Engine generates two scores from the collected data every session. The Cognitive Health Index (CHI) combines MMSE/MoCA scores from dialogue, latency, orientation, and pet interaction scores into a single 0–100 scale with confidence intervals. The Emotional Well-being Index (EWI) combines sentiment scores, prosody, mood probes, and pet engagement scores. Both indexes are presented on the Caregiver Dashboard as trends and can be set to generate alerts at configurable threshold levels.

5 Data Privacy and Ethical Considerations

Given the sensitive nature of the data collected such as dialogue, cognitive performance scores, emotional indicators, family content, and behavioral logs from a highly vulnerable demographic, privacy and ethics are core considerations from the outset.

- Explicit, understandable consent must be obtained from the user (and guardian where applicable) for data collection from dialogue, pet sensors, social media accounts, and audio recordings. The robot pet's AI nature is always disclosed and consent can be withdrawn at any time without impacting care.
- Only data required for clinical analysis is collected. On-device processing is used wherever possible to minimize data transmission, while social media content is filtered and metadata is anonymized [39].
- Data is encrypted both at rest (using AES-256) and in transit (TLS 1.3). Role-based access ensures that clinical data is accessible only to authorized clinicians, family data to consenting family members, and no one else can access the chat logs. Access records are kept in an immutable audit log [12].
- To allow personalization, federated learning is used to update model weights locally without sending raw data to the server. The base LLM is hosted centrally and accessed via API endpoints [38].
- The framework respects the guidelines provided by Sharkey and Sharkey [25]: transparency, dignity, autonomy, and the understanding that robotic pets supplement, not substitute, human companionship.
- Compliance with regulations. The system is designed to comply with the EU GDPR, US HIPAA, and UAE Federal Decree-Law No. 45 of 2021 on Personal Data Protection. Clinical use requires IRB approval, and the evaluation plan in Sect. 6 takes this into account.

6 Evaluation Plan

A mixed-methods approach is adopted, beginning with a pilot study and proceeding to a more extensive controlled trial.

6.1 Quantitative Measures

- MMSE and MoCA scores generated from dialogue will be correlated with neuropsychologist-administered test scores at baseline and week six. Sensitivity, specificity, and agreement will be calculated.
- Weekly CHI scores will be analyzed using linear mixed-effect models, adjusting for baseline, age, and condition.
- EWI will be analyzed using the same approach, paying special attention to the correlation between pet interaction frequency and EWI trajectory.

- Session duration, initiation, touch duration, and time-of-day patterns will be collected automatically.
- The standard SUS will be used at weeks two and six [3].
- Baseline and follow-up measures will be taken, stratifying by condition.

6.2 Qualitative Measures

- 30–60-min interviews with 12 to 15 participants discussing empathy, emotional bonding with the pet, comfort with AI-assisted cognitive and emotional assessment, and thoughts on family data integration.
- Two focus groups with caregivers and family members on observations, dashboard usability, and ethical issues.
- Thematic analysis: Interviews will be coded using reflexive thematic analysis [6].

7 Discussion

The framework transcends the traditional dichotomy between AI and physical robots. By combining the intelligence of a large language model with a robotic pet body, it offers the benefits of rich dialogue, tactile experience, passive monitoring, and social facilitation in one unit, none of which could be achieved separately. The distinction between single-user and group modes is necessary to address the two manifestations of social isolation: lack of contact and lack of emotional connection. Single-user mode is designed to combat the former, whereas group mode combats the latter. In a care facility, patients may be in constant physical proximity, but emotionally detached. By relaying family social media content through the pet, the system helps combat a type of isolation that is difficult to address with existing clinical frameworks: exclusion from family digital communication channels. As a messenger, the pet can help maintain the user's connection with their family in a manner appropriate to their cognitive and digital literacy level. Some limitations apply. This framework is currently theoretical and not yet implemented or empirically tested. The costs of purchasing robotic pet hardware is a potential issue, although the software emulation mode may prove useful. Attitudes towards robotic companions vary across cultures, and findings from Japan or other Western countries may not be readily generalizable to the UAE context. Future research will focus on implementing the framework, conducting IRB-approved piloting studies, integrating wearable devices for more comprehensive monitoring of emotional state, and employing reinforcement learning for personalized interaction. Longer term, integration with telemedicine systems and national health infrastructure will enable clinical hand-off and wider deployment. Another feature that could be added to this model is deciding the exact animal which could help solve the problems of anxiety and depression common among old people. For example, people who own dogs have been reported to have higher self-esteem, greater physical activity and better group interaction, [17] reported that dogs especially help with social interaction and can help build an interactive community. Additionally, they can also help improve physical activity. In contrast birds are described as more suitable pets for people who demand companionship without high social demand. White et al. [36] reported that birds have been found to have a significant role in reducing stress and stimulating relaxation without demanding much social interaction, which is highly useful for introverted people. After analyzing the wellbeing index scores the model itself can suggest the most accurate animal to meet the needs of the user. This can help reduce the uncanny valley and boost personalization.

8 Conclusion

This paper has presented a framework for elderly cognitive, emotional, and social engagement that combines LLM-driven conversation with robotic pet companions. Two deployment modes, single-user companionship and group facilitation, are grounded in established work on social isolation, animal-assisted therapy, and human-robot interaction. The comparison with digital-only and humanoid alternatives shows that the approach uniquely combines tactile presence, high acceptance, passive monitoring, and continuous animal assisted style care. The architecture consists of twelve functional modules, while the monitoring pipeline uses a combination of dialogue inputs and sensor data for a multimodal assessment of well-being. The proposed evaluation strategy provides a roadmap from pilot study to longitudinal experiment. With an increasing number of aged people and limitations of clinical practice becoming apparent, scalable, ethical, and technology-driven companionship solutions will gain more importance. The proposed framework is a stepping stone towards such a future where everyone above 65 years of age could have an intelligent and empathic physical companion.

Acknowledgment. The authors gratefully acknowledge the support by the United Arab Emirates University, UAE.

References

1. Abdollahi, H., Mahoor, M.H., Zandie, R., Siewierski, J., Qualls, S.H.: Artificial emotional intelligence in socially assistive robots for older adults: a pilot study. IEEE Trans. Affect. Comput. **14**(3), 2020–2032 (2022)
2. Albasheer, O., Abdelwahab, S.I., Zaino, M.R., et al.: The impact of social isolation and loneliness on cardiovascular disease risk factors: a systematic review, meta-analysis, and bibliometric investigation. Sci. Rep. **14**(1), 12871 (2024)
3. Alshamari, M.A., Althobaiti, M.M.: Usability evaluation of wearable smartwatches using customized heuristics and system usability scale score. Future Internet **16**(6), 204 (2024)

4. Beetz, A., Uvnäs-Moberg, K., Julius, H., Kotrschal, K.: Psychosocial and psychophysiological effects of human-animal interactions: the possible role of oxytocin. Front. Psychol. **3**, 234 (2012)
5. Bernabei, S., Moretti, L., Morrone, F., Pedrinazzi, A.: Benefits of animal-assisted therapy in elderly patients: literature review. Geriatric Care **5**(1), 8027 (2019)
6. Braun, V., Clarke, V.: Using thematic analysis in psychology. Qual. Res. Psychol. **3**(2), 77–101 (2006)
7. Capiola, A., Rodeheffer, C., Hancock, B.: Age differences in uncanny valley responses to a novel social robot platform: the role of cognitive flexibility. Comput. Hum. Behav. **120**, 106762 (2021)
8. Chaturvedi, R., Verma, S., Das, R., Dwivedi, Y.K.: Social companionship with artificial intelligence: recent trends and future avenues. Technol. Forecast. Soc. Chang. **193**, 122634 (2023)
9. Chen, J., Liu, Z., Huang, X., et al.: When large language models meet personalization: perspectives of challenges and opportunities. World Wide Web **27**(4), 42 (2024)
10. Chen, M., Cao, X., Wang, A., et al.: A global perspective on risk factors for social isolation in community-dwelling older adults: a systematic review and meta-analysis. Arch. Gerontol. Geriatr. **116**, 105211 (2024)
11. Cotten, S.R., Anderson, W.A., McCullough, B.M.: The impact of ICT use on the wellbeing of older adults. Gerontologist **53**(5), 843–851 (2013)
12. Goktas, P., Grzybowski, A.: Shaping the future of healthcare: ethical clinical challenges and pathways to trustworthy AI. J. Clin. Med. **14**(5), 1605 (2025)
13. Hong, C., Tai, T., Zhou, J., et al.: Effect of home-based and remotely supervised combined exercise and cognitive intervention on older adults with mild cognitive impairment (cogito): study protocol for a randomised controlled trial. BMJ Open **14**(8), e081122 (2024)
14. Islam, R., Bae, S.W.: FacePsy: an open-source affective mobile sensing system — analyzing facial behavior and head gesture for depression detection in naturalistic settings. Proc. ACM Hum.-Comput. Interact. **8**(MHCI), 1–32 (2024)
15. Kaźmierczak, I., Jakubowska, A., Pietraszkiewicz, A.: Natural language sentiment as an indicator of depression and anxiety symptoms: a longitudinal mixed methods study. Cogn. Emot. 1–10 (2024)
16. Khan, H.T.A., Addo, K.M., Findlay, H.: Public health challenges and responses to the growing ageing populations. Public Health Challenges **3**(3), e213 (2024)
17. McConnell, A.R., Brown, C.M., Shoda, T.M., Stayton, L.E., Martin, C.E.: Friends with benefits: on the positive consequences of pet ownership. J. Pers. Soc. Psychol. **101**(6), 1239 (2011)
18. Miura, C., Chen, S., Saiki, S., Nakamura, M., Yasuda, K.: Assisting personalized healthcare of elderly people: developing a rule-based virtual caregiver system using mobile chatbot. Sensors **22**(10), 3829 (2022)
19. Mori, M.: The uncanny valley. IEEE Robot. Autom. Mag. **19**(2), 98–100 (2012), original work in Energy, 7(4), 1970
20. Obler, L.K., Albert, M.L.: Language and aging: a neurobehavioral analysis. In: Beasley, D.S., Davis, G.A. (eds.) Aging: Communication Processes and Disorders, pp. 107–121. Grune & Stratton (1981)
21. Pandya, A., Lodha, P., Gupta, A.: Technology for early detection and diagnosis of mental disorders: an evidence synthesis. In: Digital Healthcare in Asia and Gulf Region for Healthy Aging and More Inclusive Societies, pp. 37–54. Elsevier (2024)
22. Raile, P.: The usefulness of ChatGPT for psychotherapists and patients. Humanit. Soc. Sci. Commun. **11**(1), 1–8 (2024)

23. Sabanovic, S., Bennett, C.C., Chang, W.L., Huber, L.: PARO robot affects diverse interaction modalities in group sensory therapy for older adults with dementia. In: Proceedings of the IEEE International Conference on Rehabilitation Robotics (ICORR), pp. 1–6 (2013)
24. Sanderson, K.: GPT-4 is here: what scientists think. Nature **615**(7954), 773 (2023)
25. Sharkey, A., Sharkey, N.: Granny and the robots: ethical issues in robot care for the elderly. Ethics Inf. Technol. **14**(1), 27–40 (2012)
26. Shibata, T., Wada, K.: Robot therapy: a new approach for mental healthcare of the elderly – a mini-review. Gerontechnology **9**(2), 129–145 (2010)
27. Siricharoen, N.: Creative brain training apps and games can help improve memory, cognitive abilities, and promote good mental health for the elderly. EAI Endorsed Trans. Pervasive Health Technol. (2023)
28. Spreadborough, K.: Emotional tones and emotional texts: a new approach to analyzing the voice in popular vocal song. Music Theory Online **28**(2) (2022)
29. Sung, J.Y., Grinter, R.E., Christensen, H.I.: Pimp my roomba: designing for personalization. In: Proceedings of the SIGCHI Conference on Human Factors in Computing Systems (CHI), pp. 193–196 (2009)
30. Tamura, T., Yonemitsu, S., Itoh, A., et al.: Is an entertainment robot useful in the care of elderly people with severe dementia? J. Gerontol. A Biol. Sci. Med. Sci. **59**(1), 83–85 (2004)
31. Tapus, A., Mataric, M.J., Scassellati, B.: Socially assistive robotics – the grand challenges in helping humans through social interaction. IEEE Robot. Autom. Mag. **14**(1), 35–42 (2007)
32. Touvron, H., Martin, L., Stone, K.: LLaMA 2: open foundation and fine-tuned chat models. arXiv preprint arXiv:2307.09288 (2023)
33. United Nations, Department of Economic and Social Affairs, Population Division: World population ageing 2023: Highlights (2023)
34. Vines, J., Pritchard, G., Wright, P., Olivier, P., Brittain, K.: An age-old problem: examining the discourses of ageing in HCI and strategies for future research. ACM Trans. Accessible Comput. **7**(1), 1–32 (2015)
35. Wada, K., Shibata, T.: Living with seal robots – its sociopsychological and physiological influences on the elderly at a care house. IEEE Trans. Neural Syst. Rehabil. Eng. **14**(4), 557–563 (2006)
36. White, M.E., Hamlin, I., Butler, C.W., Richardson, M.: The joy of birds: the effect of rating for joy or counting garden bird species on wellbeing, anxiety, and nature connection. Urban Ecosyst. **26**(3), 755–765 (2023)
37. Xygkou, A., et al.: MindTalker: navigating the complexities of AI-enhanced social engagement for people with early-stage dementia. In: Proceedings of the CHI Conference on Human Factors in Computing Systems, pp. 1–15 (2024)
38. Zhang, C., Xie, Y., Bai, H., Yu, B., Li, W., Gao, Y.: A survey on federated learning. Knowl.-Based Syst. **216**, 106775 (2021)
39. Zuo, Z., Watson, M., Budgen, D., et al.: Data anonymization for pervasive health care: systematic literature mapping study. JMIR Med. Inform. **9**(10), e29871 (2021)

Beyond Physical Safety: Vision-Language Models for Emotional Safety Reasoning in Dementia Care

Tanmayi Inaparthy and Fengpei Yuan(✉)

Worcester Polytechnic Institute, Worcester, MA, USA
{tinaparthy,fyuan3}@wpi.edu

Abstract. People living with dementia (PLWDs) often express emotion in ways that differ from cognitively healthy individuals, requiring assistive AI systems that can accurately interpret affect and respond appropriately in context. We present ***EmoTrace***, a framework for developing emotionally intelligent assistive AI agents for dementia care by fine-tuning vision-language models (VLMs) on dementia-care interactions. Building upon LongVILA, we design a multimodal pipeline that processes dementia care videos through segmentation into short clips, caption generation using NVILA-8B, transcript extraction using Whisper, and emotional reasoning-focused Question-Reasoning-Answer (QRA) generation with DeepSeek-R1. Combined with chain-of-thought supervised fine-tuning, this pipeline enables joint temporal, spatial, and affective reasoning tailored to dementia care dynamics. Across representative scenarios, the fine-tuned model produces more accurate interpretations of PLWD emotional states, more grounded attributions of caregiver influence on affect, and more contextually appropriate emotional-safety assessments than the base model. *EmoTrace* addresses a key gap in assistive AI for aging: the integration of long-context multimodal perception with dementia-specific reasoning, taking a step toward assistive systems that are not only physically safe but emotionally aware.

Keywords: vision-language models · dementia care · emotional safety · multimodal reasoning · assistive AI · multimodal AI · chain-of-thought fine-tuning · person-centered care · long-context video understanding

1 Introduction

Dementia affects over 55 million people worldwide, with projections indicating this number will triple by 2050 [33]. As prevalence rises, demand for personalized, continuous care far exceeds the availability of trained caregivers. Assistive AI technologies—smart-home monitoring, telehealth platforms, ambient assisted living systems, video-based caregiving tools, and socially interactive agents—are increasingly proposed to augment human caregiving capacity. However, these

S. S. Khan et al. (Eds.): IJCAI 2026, CCIS 3127, pp. 107–122, 2027.
https://doi.org/10.1007/978-981-92-5190-2_8

systems frequently struggle to interpret, understand, and respond appropriately to the unique emotional and behavioral expressions of people living with dementia (PLWDs), where misalignment can escalate distress and undermine care quality.

Why Dementia Changes the Affect-Recognition Problem? Compared to cognitively healthy older adults, PLWDs may display atypical affective signals, non-standard communication patterns, and heightened sensitivity to contextual stressors [4]. These differences arise from cognitive impairments affecting memory, language processing, and emotional regulation. As dementia progresses, PLWDs rely increasingly on nonverbal communication—facial expressions, gestures, body language, and vocalizations—to convey needs and emotional states [10,18]. General-purpose affect-recognition pipelines trained on typical populations therefore tend to misinterpret PLWDs' cues or miss early signs of mounting frustration, anxiety, or fear, leading to avoidable agitation or withdrawal during real interactions.

Emotional Safety as a Safety Requirement in Dementia Care. Assistive AI and connected health systems have made substantial progress on *physical safety* (e.g., fall detection, collision avoidance, hazard monitoring) [12]. Yet for PLWDs, physical safety alone is insufficient: a system can be physically safe while still causing harm if it fails to recognize distress or responds in ways that amplify confusion and agitation. We use the term ***emotional safety*** to describe a system's capacity to (1) recognize and interpret a person's affective state from multimodal cues, and (2) respond in ways that reduce psychological distress rather than exacerbate it. Emotional safety is particularly critical in dementia care because behavioral and psychological symptoms of dementia (BPSD)—including agitation, anxiety, depression, apathy, and irritability—are prevalent and strongly associated with caregiver burden, institutionalization, and diminished quality of life [6,16]. Moreover, PLWDs often cannot verbally articulate their emotional states; their distress manifests through facial expressions, body posture changes, repetitive gestures, and vocal tone shifts that require nuanced multimodal interpretation [4,18]. Furthermore, inappropriate or emotionally misaligned interactions during caregiving—for example, a rushed tone during dressing assistance or persisting with a task without acknowledging mounting frustration—can trigger catastrophic emotional reactions including combativeness, prolonged crying, or complete withdrawal [4,16].

Why Current Assistive AI Falls Short? Most existing systems often emphasize *task completion* (e.g., was the medication reminder delivered? did the system guide the PLWD to the dining room?), treating PLWDs as passive recipients of services rather than as individuals with dynamic emotional landscapes that must be continuously monitored and respected. Many also rely on unimodal or weakly coupled sensing pipelines that analyze facial expressions, speech, or activity in isolation. Such fragmented perception limits their ability to understand how caregiver actions and environmental factors *jointly* shape affect over time. When technology observes a PLWD through cameras, microphones, and ambient sensors but cannot *integrate* visual behavior, linguistic interaction, and temporal context into a coherent understanding of emotional state, we risk building systems that are physically safe but emotionally harmful.

Our Position. Vision-language models (VLMs), when fine-tuned for dementia-specific emotional reasoning, can provide a foundational perceptual and reasoning layer for emotionally safe assistive AI. By jointly encoding video observations, language (e.g., transcripts or caregiver speech), and temporally-grounded structured reasoning supervision within a unified architecture, VLMs can integrate facial expressions, gestures, posture, interaction context, and their evolution over time [13]. Crucially, this multi-modal capability is not tied to a single deployment platform: the same fine-tuned backbone can support smart-home systems that detect distress in PLWD living alone, telehealth platforms that augment clinician assessments by providing automated emotional state tracking during remote consultations [28], caregiver-training tools that analyze recorded care interactions, and ambient assisted living environments can adapt lighting, music, and activity suggestions to inferred affect [30].

Contributions: In this work, we introduce **EmoTrace**, a framework for emotionally safe assistive AI built on domain-adapted fine-tuning of long-context vision-language models. Our key contributions are:

- We formalize *emotional safety* as a core design requirement for assistive AI in dementia care, emphasizing the need for integrated interpretation of affect, behavior, and interaction context over time.
- We propose a scalable multimodal data pipeline that aligns long-form caregiving videos with speech transcripts and converts them into structured Question–Reasoning–Answer (QRA) supervision targeting emotional trajectories, caregiver influence, and safety-relevant behaviors.
- We develop a reasoning-centered fine-tuning strategy that adapts long-context VLMs using chain-of-thought supervision on domain-specific QRA data, enabling integrated perception and interpretation of facial expressions, gestures, posture, speech, and temporal dynamics.
- Through training-dynamics analysis and qualitative reasoning comparison, we demonstrate that EmoTrace produces more detailed, context-aware, and clinically meaningful emotional reasoning than its base VLMs, particularly in scenarios involving escalation, de-escalation, and multimodal ambiguity.

2 Related Work

Our work intersects three active research areas: video reasoning with vision-language models, multimodal emotion understanding with large models, and AI for dementia care and assisted living.

Vision-Language Models for Long-Form Video Reasoning. Recent VLMs have transformed video understanding from short-clip classification into a unified multimodal reasoning paradigm [31]. Early VideoLLM systems such as Video-ChatGPT [22] extended image-based VLMs to short segments but offered little support for modeling evolving interpersonal dynamics. Subsequent systems introduced hierarchical and multi-resolution architectures: LongVLM [32] decomposes videos into event-level segments, MovieChat [29] adds memory consolidation for multi-turn dialogue, and HierarQ [3] further improves efficiency through hierarchical query modeling. LongVILA [8] introduced Multi-Modal Sequence Parallelism to support thousands of frames per sample, and its successor LongVILA-R1 [7] combines chain-of-thought supervised fine-tuning with reinforcement learning to enhance multi-step reasoning. Despite this progress, all of these systems are trained on entertainment datasets [35] and lack domain-specific supervision for dementia care. **EmoTrace** targets this gap by fine-tuning a long-video VLM with multimodal QRA supervision derived from dementia caregiving videos.

Multimodal Emotional Understanding with Large Models. Multimodal large language models has shifted emotion recognition toward unified architectures that jointly encode visual, linguistic, and acoustic information [34]. Research [28] also shows that incorporating facial cues improves the responses of mental-health chatbots. FaceLLM [24] and FacexBench [23] establish evaluation protocols for face-centric multimodal reasoning. Despite these advances, most emotion recognition systems still focus on *label prediction* rather than *interaction-level reasoning*, and their training data overwhelmingly reflects cognitively healthy populations [18]. **EmoTrace** differs by targeting context-aware emotional reasoning grounded in dementia caregiving interactions.

AI for Dementia Care and Emotional Safety. AI applications in dementia care have traditionally emphasized diagnosis and monitoring [2], with video-based behavioral analysis applied to detect agitation [17], and pain [20]. Although physical safety remains a foundational requirement of assistive systems [12], clinical evidence consistently indicates that physical safety alone is insufficient in dementia care [6,14]. Affective technologies for PLWDs–including socially assistive agents such as Ryan [1], and engagement-assessment frameworks such as ENGAGE-DEM [25] – have shown the value of integrating affective perception with dialogue management, and meta-analyses report reductions in depression and anxiety among PLWDs interacting with such systems [14]. However, most existing systems rely on modular pipelines and shallow classifiers, struggling to interpret complex interaction dynamics. Ethical scholars further emphasize that personhood and agency persist under cognitive impairment, with value-driven design frameworks [5] and critiques of dignity erosion [27] highlighting the importance of meaningful emotional interpretation. **EmoTrace** departs from prior paradigms by treating dementia video understanding as a multi-modal *reasoning* problem, leveraging VLMs fine-tuned with structured QRA supervision to integrate behavior, language, and situational context in line with the ethical and relational demands of dementia care.

3 Methodology

Overview. Our *EmoTrace* framework consists of five main stages: (1) video segmentation, (2) caption generation, (3) caption merging, (4) QRA generation, and (5) supervised fine-tuning (SFT). Figures 1 and 2 illustrate the complete pipeline, which is designed as a reusable component for assistive-AI applications (e.g., smart-home monitoring, telehealth, and caregiver-training tools).

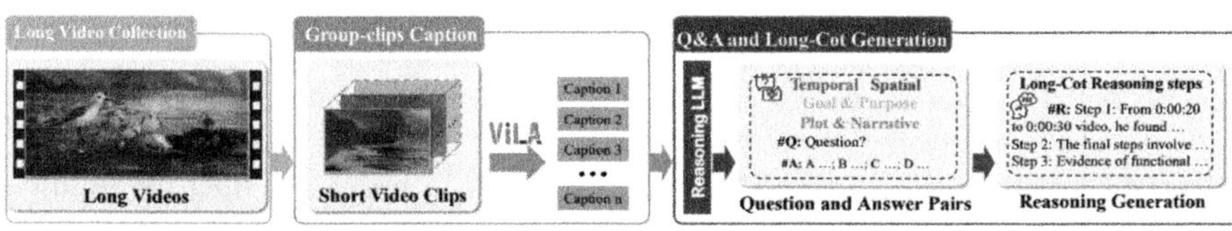

Fig. 1. General pipeline template adapted from LongVILA-R1 [7]. Long videos are segmented into short clips, captioned, merged into chronological narratives, and converted into QRA triplets that drive chain-of-thought supervised fine-tuning. *EmoTrace* specializes this template for dementia care (see Fig. 2).

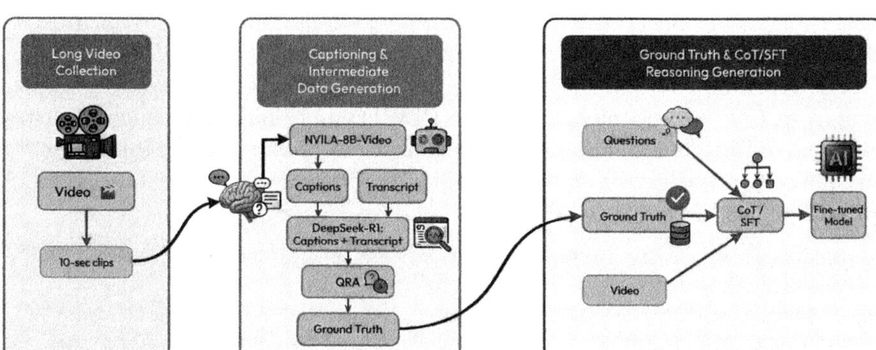

Fig. 2. *EmoTrace* data and training pipeline. Dementia care videos (left) are segmented into 10-second clips and processed through NVILA-8B-Video captioning and Whisper-based transcription (middle). DeepSeek-R1 then converts aligned caption–transcript pairs into chain-of-thought QRA supervision used to fine-tune LongVILA-R1-7B (right), yielding a model that infers PLWD affect and reasons about how caregiver actions and environmental factors shape it over time.

3.1 Video Dataset

AI-assisted dementia caregiving face a critical bottleneck: the scarcity of publicly available training data. To address this constraint, we leveraged publicly available educational content from verified healthcare institutions on YouTube, focusing on dementia care scenarios that mirror real-world caregiving challenges. We systematically collected videos using targeted search queries including "dementia care," "Alzheimer's caregiving," "dementia symptoms," and "caregiver training," prioritizing content from established medical institutions and healthcare education channels, including UCLA Alzheimer's and Dementia Care Program and CareAcademy. Using yt-dlp, we downloaded 42 videos totaling 3 h and 32 min of high-quality caregiving demonstrations. The videos span diverse dementia care scenarios including agitation management, communication strategies, behavioral challenges (wandering, repetitive behaviors, refusal to bathe), and activities of daily living.

Video Segmentation. Raw dementia care videos vary in length from several minutes to over an hour. To enable fine-grained analysis while managing computational constraints, we segment videos into 10-second clips. To prevent data leakage between adjacent clips, splits were performed at the video level, ensuring that all 10-second clips derived from a given video appear exclusively in either the training or validation set, never both.

3.2 Multimodal Data Pipeline

Caption Generation. For each 10-second clip, we generate detailed visual descriptions using NVILA-8B-Video [21], a vision-language model optimized for video captioning. Our prompting strategy emphasizes emotional and behavioral analysis relevant to PLWD care, as illustrated as below. Generated captions range from 150–300 tokens and capture PLWD emotional states such as distress, confusion, agitation, and calmness; caregiver behaviors including tone indicators, body language, and intervention strategies; environmental context; and the temporal dynamics of emotional escalation and de-escalation.

Captioning Prompt

"Describe this dementia care video clip in detail. Focus on: 1. Their apparent emotional state at the start of the clip, 2. How their emotions change as the scene unfolds, 3. Visual cues (gestures, posture, facial movements), 4. Why they react the way they do in each moment."

Caption Merging and Timeline Construction. Individual clip captions are chronologically merged to create comprehensive video narratives. This step is crucial for enabling temporal reasoning, as many dementia care scenarios involve behavioral patterns that unfold over minutes rather than seconds. When merged, we ensure the timestamped nature of data is retained.

QRA Generation via DeepSeek-R1. We employ DeepSeek-R1 [11], a reasoning-focused large language model, to generate Question–Reasoning–Answer (QRA) triplets from merged visual captions and speech transcripts. We first align timestamped captions with overlapping transcript segments, producing unified multimodal timelines for each video.

For each video, we automatically generate multiple reasoning-focused questions using a curated set of thematic templates covering: (1) identification of primary and secondary emotional states from facial expression and posture; (2) temporal shifts in affect and behavioral escalation; (3) caregiver actions that unintentionally intensify or alleviate distress; (4) environmental and social triggers of agitation; (5) early warning signs preceding emotional breakdown; (6) hidden or masked emotions revealed through subtle cues; and (7) effectiveness of de-escalation strategies. Our prompt emphasizes temporal reasoning, safety assessment, and multi-step inference:

> *"Based on the following dementia care video captions and transcripts, generate challenging multiple-choice questions that require reasoning about {theme}. For each question, provide: the question, four multiple-choice options, the correct answer, and detailed chain-of-thought reasoning referencing specific timestamps."*

To improve dataset quality, we enforce uniqueness constraints by explicitly providing previously generated questions for each video, preventing semantic duplication. On average, we generate 40 QRA pairs per video, yielding a total of about 1,600 training examples. Figure 3 presents a representative example.

Data Formatting and Preprocessing. Each QRA triplet is serialized into a JSON record containing the video path, question, chain-of-thought reasoning enclosed in <think> tags, and final answer in <answer> tags. During preprocessing, videos are loaded and sampled to 32 frames at 336×336 resolution. Text is tokenized, with question tokens masked and only reasoning and answer tokens included in loss computation during training.

3.3 Training Pipeline

Supervised Fine-Tuning (Chain-of-Thought SFT). Our training pipeline employs supervised fine-tuning on the custom QRA dataset using Low-Rank Adaptation (LoRA) [15] to efficiently adapt LongVILA-R1-7B. Rather than updating all 7.24B parameters, we insert lightweight low-rank adapter matrices into attention projection layers, enabling task-specific adaptation while preserving the pretrained backbone. LoRA is applied to both the language backbone and the SigLIP vision encoder, allowing joint refinement of visual perception and multimodal reasoning. Table 1 summarizes the configuration.

Question: A caregiver in the video is shown organizing a closet and reducing clothing choices while the text "Simplify environment. Reduce choices" is displayed. What is the primary intention behind this demonstrated strategy?

Options:
(A) To minimize laundry costs by limiting wardrobe options
(B) To accelerate dressing routines for caregiver efficiency
(C) To decrease decision-making stress and agitation in PLWDs
(D) To encourage PLWD independence in complex tasks

Answer: (C)

Reasoning: The video emphasizes person-centered care principles where environmental modifications aim to reduce cognitive load on PLWDs. At timestamp 120–130s, the caregiver deliberately simplifies the closet while on-screen text reinforces the principle of reducing choices. This aligns with evidence-based dementia care practices that recognize decision-making tasks can trigger agitation in cognitively impaired individuals. Options A and B reflect efficiency-focused perspectives that contradict the patient-centered philosophy. Option D is incorrect because increasing independence in complex tasks would require more, not fewer, choices.

Fig. 3. A representative multimodal QRA triplet generated by DeepSeek-R1 from a caregiver-PLWD video. The reasoning step grounds the answer in a specific timestamped event (120–130 s) and contrasts each distractor with evidence-based person-centered care principles.

Table 1. LoRA configuration for supervised fine-tuning.

Parameter	Value
Rank (r)	16
Scaling factor (α)	32
α/r	2.0
Target layers	Query, Value projections
Applied modules	LLM + SigLIP vision tower
LoRA dropout	0.05
Trainable parameters	$\sim$50M ($<$1%)

Low-Resource Prototyping with QLoRA. During early experimentation under single-GPU constraints, we adopted QLoRA [9], training LoRA adapters on top of a 4-bit NF4-quantized backbone with double quantization enabled. Base model weights remain frozen in 4-bit precision while LoRA adapter weights are maintained in bfloat16. This reduced memory from approximately 65 GB (full-precision LoRA) to roughly 18 GB on a single 80 GB NVIDIA H100, enabling rapid prototyping.

Training Objective and Regularization. Our training objective follows the chain-of-thought supervised fine-tuning paradigm introduced in LongVILA-R1 [7]. The model generates structured intermediate reasoning enclosed within <think> tags, followed by final answer tokens within <answer> tags. Question tokens are masked from the loss computation. Given the dataset size of approximately 1,600 QRA samples, we applied early stopping with a patience of three epochs. Best-performing checkpoints consistently emerged between epochs 8–12, aligning with prior observations in parameter-efficient fine-tuning literature.

Implementation Details. The training pipeline was implemented in PyTorch 2.1 with HuggingFace Transformers 4.38. We incorporated lazy video decoding, robust dataset patching mechanisms, and static graph optimizations. Long-context video processing introduced memory bottlenecks in the SigLIP vision encoder due to quadratic self-attention complexity. We addressed this with chunked frame encoding and activation checkpointing, reducing peak memory consumption by approximately 40% relative to naive full-batch encoding. Training used AdamW with a learning rate of 2×10^{-4}, cosine decay scheduler, 3% warmup, weight decay of 0.01, and gradient clipping at 1.0. We used a per-device batch size of 1 with gradient accumulation over 4 steps (global batch size 16), bfloat16 precision, up to 120 frames per clip, and a maximum context length of 32,768 tokens.

3.4 Multimodal Enhancement

To augment visual information with audio context, we integrate speech transcripts using Whisper [26]. This multimodal integration allows the model to assess caregiver tone (rushed, calm, or empathetic), categorize PLWD verbal responses (resistant, confused, or accepting), and identify instances where verbal and visual cues misalign—such as when a caregiver says "Don't worry" while displaying frustrated body language [16]. DeepSeek is also given the preliminary task of sorting speaker roles based on context.

4 Results and Discussion

We evaluate *EmoTrace* through combined analysis of training dynamics and qualitative reasoning behavior, comparing the base LongVILA-R1-7B model with the fine-tuned EmoTrace variant across representative caregiving interactions involving emotional escalation, de-escalation, and real-world multimodal integration.

Training Dynamics and Convergence. Figure 4 illustrates training loss across fine-tuning epochs, with validation loss monitored separately to guide early stopping. The model exhibits rapid loss reduction during early training stages, indicating effective adaptation to domain-specific supervision. Validation

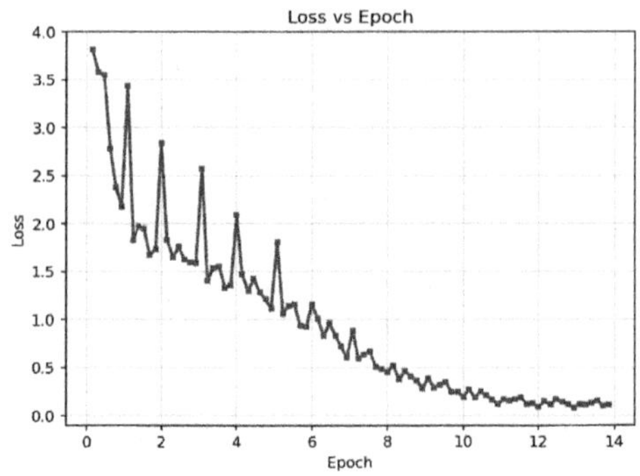

Fig. 4. Training loss curve for LongVILA-R1-7B fine-tuned with QLoRA on 4×H100 GPUs, aggregated by epoch. The steady downward trend indicates stable optimization and convergence over the course of training. Validation loss is monitored separately to determine early stopping but is not plotted here.

loss stabilizes between epochs 8 and 12, reflecting the emergence of robust emotional reasoning patterns. Beyond epoch 15, validation loss begins to increase while training loss continues to decrease, signaling overfitting. This motivated early stopping and the selection of intermediate checkpoints (epochs 8–10) for evaluation.

Qualitative Reasoning Comparison. Figure 5 presents a qualitative comparison on a representative dementia care interaction in which the PLWD exhibits visible distress and resistance during a caregiving task. The base model interprets the caregiver's physical proximity and touch as inherently supportive, concluding that agitation is resolved through caregiver support despite observable resistance and sustained distress. This reasoning reflects reliance on generic caregiving heuristics and underweights behavioral signals such as pulling away, defensive posture, and continued frustration.

In contrast, the EmoTrace model integrates resistance cues into its interpretation. It explicitly references behaviors such as pulling clothing away, appearing frustrated, and maintaining a closed or defensive posture, linking these to loss of autonomy and caregiver persistence. The model identifies that task-focused actions without emotional validation can exacerbate distress, correctly classifying the interaction as emotionally escalatory.

In de-escalation scenarios, both models select the correct answer; however, their reasoning quality differs substantially. While the base model provides high-level descriptions of reassurance, EmoTrace traces specific behavioral transitions over time—noting shifts from head-holding and defensive posture to relaxed sitting and cooperative engagement—and connects these changes to concrete interventions such as hand-holding, guided positioning, and sustained supportive presence.

Question: How can we describe the emotional state of the person with dementia in this clip, and how did the caregiver's actions affect that emotional state?
Options:
(A) Calm and cooperative, caregiver maintained that state.
(B) Agitated and resistant, caregiver escalated feelings.
(C) Fearful and frustrated, caregiver helped her feel safer.
(D) Indifferent, caregiver had no impact.
Base Model Answer: **(C)**
Reasoning: <think> The person with dementia appears to be in a state of agitation, as indicated by their body language, such as holding their head and looking distressed. The caregiver's actions, such as gently placing a hand on their back and offering support, suggest an attempt to calm them. This interaction is consistent with caregiving practices aimed at reducing anxiety and providing reassurance. The caregiver's presence and actions seem to help the person with dementia feel more secure, as they are seen standing close by, which is typical in caregiving scenarios to manage distress. **The caregiver's actions are not dismissive or confrontational, but rather supportive and attentive, which helps to reduce the person's agitation** The caregiver's consistent presence and gentle gestures indicate a deliberate effort to create a calming environment, which is effective in managing the emotional state of a person with dementia. This approach aligns with the goal of making the person feel safer and more cooperative <think>
Fine-Tuned Model Answer: (B)
Reasoning: <think>The person with dementia is shown in a state of distress, holding their head and appearing agitated. This indicates a high level of anxiety or confusion, which is a common emotional state in dementia. The caregiver's initial approach is to assist the person with putting on a plaid shirt, which is a common, non-threatening task. **However, the person resists, pulling the shirt away and appearing frustrated** This resistance is a common behavioral response in dementia, often due to a loss of autonomy and a desire to maintain control. The caregiver's response is to persist in the task, which can be overwhelming and stressful for the person with dementia, potentially exacerbating their anxiety.
The caregiver's actions throughout the video are focused on completing tasks, such as putting on a shirt and holding a blanket, without addressing the person's emotional state This approach does not provide comfort or reduce anxiety, which are critical in managing dementia-related distress. The person's body language consistently shows signs of agitation and frustration, indicating that the caregiver's actions did not help to alleviate these feelings but instead likely contributed to them <think>

Fig. 5. Qualitative comparison on a dressing-assistance scenario. Both models correctly identify the PLWD as fearful and frustrated, but only *EmoTrace* attributes the emotional escalation to task-focused caregiver actions (e.g., urging shirt completion without acknowledging distress). Highlighted text marks key reasoning differences.

Multimodal Reasoning with Transcript Integration. Figure 6 presents a qualitative comparison when both visual and transcript information are available. The base model classifies the interaction as emotionally escalatory (Option B), focusing primarily on early visual cues and treating them as persistent indicators of distress. In contrast, EmoTrace integrates visual observations with linguistic information from subtitles and transcripts, identifies a shift in caregiver behavior from confrontational to calm, and links this change to corresponding improvements in the patient's posture and engagement. By referencing subtitles describing the approach as "gentle and supportive," the model recognizes successful de-escalation and correctly selects Option C. This example demonstrates EmoTrace's ability to revise its emotional interpretation as new multimodal evidence becomes available. Rather than relying solely on initial visual impressions, the fine-tuned model tracks interaction dynamics over time and reconciles visual and linguistic cues into a coherent emotional narrative.

Question: How can we describe the emotional state of the person with dementia in this clip, and how did the caregiver's actions affect that emotional state?

Options:
(A) Calm and cooperative.
(B) Agitated and resistant, caregiver escalated feelings.
(C) Fearful and frustrated, caregiver helped her feel safer.
(D) Indifferent.

Base Model — Answer: (B)
Reasoning:
<think>The person with dementia appears agitated, as indicated by her raised hand and tense posture. The caregiver's confrontational stance and assertive gestures likely increased her anxiety. The caregiver's rigid approach suggests a lack of flexibility and respect for autonomy, which further escalates distress.</think>

Fine-Tuned Model + Transcript — Answer: (C)
Reasoning:
<think>The person initially appears agitated, lying defensively on the bed with distressed posture. The caregiver's early confrontational stance contributes to escalation. However, the interaction shifts when the caregiver sits beside the person and adopts a calm, attentive demeanor. The person's posture changes to sitting upright and engaging with a book, indicating reduced agitation. Subtitles describing the approach as "gentle and supportive" reinforce this transition. The coordinated visual and linguistic cues demonstrate successful de-escalation.</think>

Fig. 6. Multimodal reasoning when both visual frames and transcripts are available. The base model anchors on the initial agitated posture and selects (B) escalation; EmoTrace integrates the caregiver's shift in tone (from subtitles) with corresponding changes in the PLWD's posture and engagement, revising the assessment to (C) effective de-escalation.

Summary of Findings. Across all evaluated scenarios, EmoTrace demonstrates stronger alignment between visual cues, transcript context, and emotional inference. It more reliably associates facial tension, gaze aversion, and withdrawal gestures with confusion or distress, and integrates caregiver speech and actions into coherent emotional narratives. Where the base model relies on generalized caregiving heuristics, EmoTrace applies dementia-specific interpretive frameworks learned through structured QRA supervision.

These findings support our central thesis: while long-context VLMs possess the representational capacity to process multimodal and temporal information, domain-adapted fine-tuning is necessary to transform this capacity into emotionally safe reasoning.

4.1 Generality of the *EmoTrace* Pipeline

While we validate *EmoTrace* in the dementia care domain, the underlying pipeline constitutes a general methodology for building emotionally safe AI systems. For example, in *general elder care*, cognitively healthy elderly individuals may have difficulty expressing needs due to sensory impairments or social isolation [36]. In *rehabilitation*, patients recovering from stroke or traumatic brain injury may experience frustration or pain not verbally reported but visible through multimodal cues [19]. In *mental health monitoring*, patients with depression, PTSD, or psychosis exhibit subtle behavioral changes that unfold over time and require longitudinal multimodal analysis. In *connected caregiving technologies* such as smart-home monitoring, telehealth platforms, and ambient assisted living, similar architectures could enhance emotional safety by enabling continuous, context-aware interpretation of resident or patient affect. In each of these domains, the core challenge is identical: technology that observes humans must *understand* what it observes in emotional terms. The *EmoTrace* pipeline provides a reproducible, domain-adaptable methodology for achieving this through VLM fine-tuning.

4.2 Limitations and Future Work

Several limitations of the present study point to clear next steps. (1) Our dataset, while carefully curated, is limited in size due to the sensitive nature of dementia care videos; scaling through partnerships with care facilities and synthetic data augmentation is an active priority. Additionally, the dataset consists of educational YouTube videos which likely feature actors or trained caregivers enacting curated scenarios; while these actors endeavor to authentically recreate real patient behavior, such staged representations may still fail to capture the spontaneous, unpredictable, and subtle behavioral nuances of genuine dementia patients in real-world care settings, and future work should validate on clinical recordings. (2) We report training-dynamics and qualitative analyses but not a full quantitative evaluation, so a natural next step is to measure multiple-choice accuracy and chain-of-thought faithfulness, and to quantitatively compare against additional baselines such as the base LongVILA-R1 [7] and general-purpose multimodal LLMs without dementia-specific adaptation. (3) Although our QRA supervision is grounded in publicly available caregiving content, the QRA triplets are synthetically generated by DeepSeek-R1 without human annotation, meaning the fine-tuned model may risk inheriting DeepSeek-R1's interpretive biases; to mitigate this, we plan to engage geriatricians and dementia-care nurses in structured rating studies to clinically validate DeepSeek-R1's generated reasoning, assess clinical accuracy, the appropriateness of caregiver-influence attributions, and potential safety risks of model outputs in realistic care contexts. (4) Our pipeline currently uses only the supervised fine-tuning stage of the LongVILA-R1 recipe; extending it with the reinforcement-learning stage from the original LongVILA-R1 work [7]—using reward signals informed by clinician feedback and answer correctness—is a natural next step toward improving reasoning robustness, reducing hallucinated cues, and supporting real-time, privacy-preserving deployment in real care settings.

5 Conclusion

We introduced EmoTrace, a framework that adapts long-context vision-language models for emotionally informed dementia care by aligning video, transcripts, and structured Question–Reasoning–Answer supervision. Across representative scenarios, the fine-tuned model produces more reliable associations between visual cues and underlying affect, distinguishes escalation from de-escalation patterns, and generates temporally grounded explanations that reflect real caregiving conditions more detailed than the base model. More broadly, this work positions emotional safety as a core requirement for assistive AI in dementia care: systems that merely avoid physical harm are insufficient if they misinterpret nonverbal distress signals or fail to integrate affect with interaction context. While demonstrated in dementia care, the underlying methodology offers a potential blueprint for multimodal emotion understanding in other high-sensitivity domains such as mental-health support, rehabilitation, and connected health. Wherever technology must interpret human distress and emotion, domain-adapted emotionally-safe multimodal VLM architectures such as EmoTrace suggest a promising new direction for human-centered AI.

References

1. Abdollahi, H., Mahoor, M.H., Zandie, R., Siewierski, J., Qualls, S.H.: Artificial emotional intelligence in socially assistive robots for older adults: a pilot study. IEEE Trans. Affect. Comput. **14**(3), 2020–2032 (2023). https://doi.org/10.1109/TAFFC.2022.3143803
2. Aditya Shastry, K., Sanjay, H.: Artificial intelligence techniques for the effective diagnosis of alzheimer's disease: a review. Multimedia Tools Appl. **83**(13), 40057–40092 (2024)
3. Azad, S., Vineet, V., Rawat, Y.S.: HierarQ: task-aware hierarchical Q-Former for enhanced video understanding. In: Proceedings of the IEEE/CVF Conference on Computer Vision and Pattern Recognition (CVPR), pp. 8545–8556 (2025). https://doi.org/10.1109/CVPR52734.2025.00799
4. Bender, E.N., Savundranayagam, M.Y., Murray, L., Orange, J.B.: Supportive strategies for nonverbal communication with persons living with dementia: a scoping review. Int. J. Nurs. Stud. **136**, 104365 (2022). https://doi.org/10.1016/j.ijnurstu.2022.104365
5. Buhr, E., Welsch, J., Shaukat, M.S.: Value preference profiles and ethical compliance quantification: a new approach for ethics by design in technology-assisted dementia care. AI Soc. (2024). https://doi.org/10.1007/s00146-024-01947-7

6. Cerejeira, J., Lagarto, L., Mukaetova-Ladinska, E.B.: Behavioral and psychological symptoms of dementia. Front. Neurol. **3**, 73 (2012). https://doi.org/10.3389/fneur.2012.00073
7. Chen, Y., et al.: Scaling RL to long videos. In: Advances in Neural Information Processing Systems (2025)
8. Chen, Y., et al.: LongVILA: scaling long-context visual language models for long videos. In: International Conference on Learning Representations, vol. 2025, pp. 18227–18246 (2025)
9. Dettmers, T., Pagnoni, A., Holtzman, A., Zettlemoyer, L.: QLoRA: efficient fine-tuning of quantized LLMs. In: Advances in Neural Information Processing Systems, vol. 36, pp. 10088–10115 (2023)
10. Ellis, M., Astell, A.: Communicating with people living with dementia who are nonverbal: the creation of adaptive interaction. PLoS ONE **12**(8), e0180395 (2017). https://doi.org/10.1371/journal.pone.0180395
11. Guo, D., et al.: DeepSeek-R1 incentivizes reasoning in LLMs through reinforcement learning. Nature **645**(8081), 633–638 (2025). https://doi.org/10.1038/s41586-025-09422-z
12. Haddadin, S., De Luca, A., Albu-Schäffer, A.: Robot collisions: a survey on detection, isolation, and identification. IEEE Trans. Rob. **33**(6), 1292–1312 (2017)
13. Hartsock, I., Rasool, G.: Vision-language models for medical report generation and visual question answering: a review. Front. Artif. Intell. **7**, 1430984 (2024). https://doi.org/10.3389/frai.2024.1430984
14. Hsieh, C.J., Li, P.S., Wang, C.H., Lin, S.L., Hsu, T.C., Tsai, C.M.: Socially assistive robots for people living with dementia in long-term facilities: a systematic review and meta-analysis of randomized controlled trials. Gerontology **69**(8), 1027–1042 (2023). https://doi.org/10.1159/000529849
15. Hu, E.J., et al.: LoRA: low-rank adaptation of large language models. In: International Conference on Learning Representations (ICLR) (2022)
16. Kales, H.C., Gitlin, L.N., Lyketsos, C.G.: Assessment and management of behavioral and psychological symptoms of dementia. BMJ **350**, h369 (2015). https://doi.org/10.1136/bmj.h369
17. Khan, S.S., Ye, B., Taati, B., Mihailidis, A.: Detecting agitation and aggression in people with dementia using sensors: a systematic review. Alzheimer's Dement. **14**(6), 824–832 (2018). https://doi.org/10.1016/j.jalz.2018.02.004
18. Khan, Z., et al.: Investigating the effects of impairment in non-verbal communication on neuropsychiatric symptoms and quality of life of people living with dementia. Alzheimer's Dement. Transl. Res. Clin. Interv. **7**(1), e12172 (2021). https://doi.org/10.1002/trc2.12172
19. Langer, A., Feingold-Polak, R., Mueller, O., Kellmeyer, P., Levy-Tzedek, S.: Trust in socially assistive robots: considerations for use in rehabilitation. Neurosci. Biobehav. Rev. **104**, 231–239 (2019). https://doi.org/10.1016/j.neubiorev.2019.07.014
20. Lautenbacher, S., Kunz, M.: Facial pain expression in dementia: a review of the experimental and clinical evidence. Curr. Alzheimer Res. **14**(5), 501–505 (2017)
21. Liu, Z., et al.: NVILA: efficient frontier visual language models. In: Proceedings of the IEEE/CVF Conference on Computer Vision and Pattern Recognition, pp. 4122–4134 (2025)
22. Maaz, M., Rasheed, H., Khan, S., Khan, F.S.: Video-ChatGPT: towards detailed video understanding via large vision and language models. In: Proceedings of the 62nd Annual Meeting of the Association for Computational Linguistics (ACL 2024), pp. 12585–12602. Association for Computational Linguistics, Bangkok, Thailand (2024)

23. Narayan, K., Vibashan, V.S., Patel, V.M.: FaceXBench: evaluating multimodal LLMs on face understanding. IEEE Trans. Biometrics Behav. Identity Sci. (2026). https://doi.org/10.1109/TBIOM.2026.3655668
24. Otroshi Shahreza, H., Marcel, S.: FaceLLM: a multimodal large language model for face understanding. In: Proceedings of the IEEE/CVF International Conference on Computer Vision Workshops (ICCVW 2025), pp. 3677–3687 (2025)
25. Perugia, G., Díaz-Boladeras, M., Català-Mallofré, A., Barakova, E.I., Rauterberg, M.: ENGAGE-DEM: a model of engagement of people with dementia. IEEE Trans. Affect. Comput. **13**(2), 926–943 (2022). https://doi.org/10.1109/TAFFC.2020.2980275
26. Radford, A., Kim, J.W., Xu, T., Brockman, G., McLeavey, C., Sutskever, I.: Robust speech recognition via large-scale weak supervision. In: Proceedings of the International Conference on Machine Learning (ICML) (2023)
27. Robinson, H., Broadbent, E., MacDonald, B.: Suitability of socially assistive robots for dementia care. J. Gerontol. **67B**(6), 699–705 (2012)
28. Schmidmaier, M., Rupp, J., Harrich, C., Mayer, S.: Using nonverbal cues in empathic multi-modal LLM-driven chatbots for mental health support. Proc. ACM Hum.-Comput. Interact. **9** (2025). https://doi.org/10.1145/3743724
29. Song, E., et al.: MovieChat: from dense token to sparse memory for long video understanding. In: Proceedings of the IEEE/CVF Conference on Computer Vision and Pattern Recognition (CVPR 2024) (2024)
30. Sorrentino, A., D'Onofrio, G., Solis, P., Mancinelli, L., Cesari, M., Ricciardi, F.: Feasibility of ambient assisted living technologies for supporting people with dementia: a systematic review. J. Alzheimers Dis. **80**(3), 1231–1248 (2021)
31. Wang, H., Liu, Z., Zhang, J., Li, X., Han, S.: Vision-language models for multimodal understanding: a survey. ACM Comput. Surv. (2024)
32. Weng, Y., Han, M., He, H., Chang, X., Zhuang, B.: LongVLM: efficient long video understanding via large language models. In: European Conference on Computer Vision (ECCV) (2024)
33. World Health Organization: Dementia (2023). https://www.who.int/news-room/fact-sheets/detail/dementia. Accessed 2025
34. Wu, C., et al.: Multimodal emotion recognition in conversations: a survey of methods, trends, challenges and prospects. In: Findings of the Association for Computational Linguistics: EMNLP 2025 (2025)
35. Wu, H., Li, D., Chen, B., Li, J.: LongVideoBench: a benchmark for long-context interleaved video-language understanding. In: Advances in Neural Information Processing Systems (NeurIPS) (2024)
36. Zhang, Y., Luo, L., Wang, X.: Aging with robots: a brief review on eldercare automation. Interdisc. Nurs. Res. **3**(1), 49–56 (2024)

AI for Rehabilitation and Clinical Decision Support

Your Voice is not (Always) You: Voice, Condition, and the Limits of Identity in Conversational UI for Aging and Rehabilitation Contexts

Naomi Unkelos-Shpigel(✉), Leon Feldman, and Mor Hodaya Maman

Department of Software Engineering, Braude College of Engineering, Karmiel, Israel
naomius@braude.ac.il

Abstract. Voice carries identity. It also carries condition: fatigue, distress, illness, neurological change. Current conversational AI architectures treat condition-related variation as noise to be suppressed in the interest of robust speaker recognition. We propose a parallel voice modeling framework in which identity and condition are learned simultaneously from the same audio-to-image representation, producing separate but co-present inferences about who is speaking and how they are. This framework is particularly relevant to aging populations, in whom vocal markers of fatigue, cognitive decline, and neurological disease progression are both more prevalent and more clinically consequential. We demonstrate the framework's feasibility on a voice dataset spanning speaker identity and emotional condition labels, reporting strong classification results across both pathways: 99.64% accuracy for speaker identity and 94.26% for vocal condition. We treat this framework not as a completed system but as a provocation: a way of making visible the design choices that current architectures have rendered invisible. If a conversational system can know not only who you are but how you are, the field faces questions about agency, consent, and the limits of identity that it has not yet been required to answer.

Keywords: Voice Analysis · Speaker Identification · Condition Detection · Parallel Modeling · Aging · Cognitive Decline · Parkinson's Disease · Privacy · Conversational AI

1 Introduction

Globally, the population aged 65 and over is projected to double by 2050, placing unprecedented strain on healthcare, rehabilitation, and assisted living systems [22]. Supporting safe, independent aging requires continuous, personalized monitoring approaches that scale beyond traditional clinical visits. Voice offers a uniquely accessible modality: it is passively available, requires no wearable sensor, and can be captured through devices already present in many homes. Yet current conversational AI systems exploit only a fraction of what the voice signal contains.

S. S. Khan et al. (Eds.): IJCAI 2026, CCIS 3127, pp. 125–137, 2027.
https://doi.org/10.1007/978-981-92-5190-2_9

Consider a common scenario in eldercare. It is 2 a.m. An older adult with early-stage Parkinson's disease is exhausted, having had disrupted sleep, a known symptom of the condition. They ask their voice assistant about their medication schedule, and the system greets them by name, authenticates them immediately, and confirms the response in its usual tone. From the system's perspective, nothing has changed. From a clinical perspective, almost everything has: vocal tremor, reduced loudness, and slowed speech rate may all signal symptom progression. This gap is not oversight. It is a consequence of how conversational AI systems are designed to listen.

Voice carries identity, the stable biometric signature that speaker verification systems have spent decades learning to extract reliably [4, 17]. But voice also carries condition: it shifts with fatigue [19], changes shape under emotional load [2, 5], degrades measurably with illness [6], and traces the arc of neurological change over months and years [8, 23]. Current architectures resolve this richness by erasure. Speaker recognition models are explicitly optimized to be robust to condition-related variation, treating fatigue, mood, and illness as noise to be filtered out [9, 14, 21]. Condition detection models, where they exist at all, operate in isolation from identity models [3, 6].

We propose a parallel voice modeling framework in which identity and condition are modeled as separable but co-present dimensions of the same vocal signal. The framework is designed with aging populations as the primary use case, where the stakes of ignoring condition information are highest. We evaluate it on a voice dataset to establish proof of feasibility, and we discuss its implications for CUI design in eldercare, rehabilitation, and chronic disease monitoring.

2 Identity and Condition in the Voice

The evidence that voice carries condition-related information is substantial and cross-disciplinary. In affective computing, spectral and prosodic features have been used to classify emotional states with increasing accuracy [15, 21, 24]. In clinical research, vocal biomarkers have been associated with depression [2, 3], fatigue and sleep deprivation [7, 19], and Parkinson's disease progression [23]. In occupational voice research, within-speaker variation across a single working day has been shown to follow recognizable patterns tied to vocal load [12]. Circadian rhythms leave measurable traces in voice [24].

For aging populations specifically, the clinical relevance of vocal condition markers is well established. Acoustic features including jitter, shimmer, harmonics-to-noise ratio (HNR), and speech rate have been validated as biomarkers for Parkinson's disease [23], with longitudinal studies showing that vocal change can precede clinical diagnosis by months [10, 20]. Vocal markers of cognitive load and fatigue are relevant to dementia monitoring [8], and depression-related vocal patterns are particularly prevalent among older adults in long-term care [2, 3]. The voice, in these contexts, is not only the medium of conversation: it is a physiological signal that CUI systems are currently discarding.

What is less well established is the relationship between the features most diagnostic for identity and those most diagnostic for condition within the same signal. Speaker verification research treats these as separable by design: the goal of i-vector [4] and x-vector [17] methods is to produce speaker embeddings invariant to condition effects. But robustness achieved by suppression is not the same as understanding. Recent work on audio-to-image encoding [1] provides a representational framework that may enable a more precise analysis by making spectral structure visible to localized attribution methods [15].

3 A Framework for Parallel Voice Modelling

We propose a parallel voice modeling framework consisting of four stages:

In the first stage, voice samples are transformed into spectral image representations using the deterministic audio-to-image encoding procedure described in [1]. The deterministic character of this step is significant: the same utterance produces the same image, meaning that differences in model behavior can be attributed to learned feature weighting rather than representational variation.

In the second stage, two separate convolutional neural networks are trained on the same encoded representations: one to classify speaker identity, one to classify vocal condition. Both networks receive the same input; they differ only in their objective and training labels. Separate networks are used deliberately rather than a multi-label architecture: independent gradient pathways ensure that Grad-CAM attribution maps for identity and condition reflect task-specific feature weighting without cross-task gradient interference, which would undermine the spatial separation analysis in Stage 3.

In the third stage, gradient-based attribution [15] generates heatmaps indicating which spectral regions were most influential for each classification decision, making it possible to examine whether identity-relevant and condition-relevant features concentrate in overlapping or distinct regions.

The fourth stage is speculative. If heatmaps reveal consistent spatial separation across speakers and conditions, this would support a deterministic analysis pathway: fixed region-based extraction producing parallel estimates without further network inference.

Figure 1 presents the model flow:

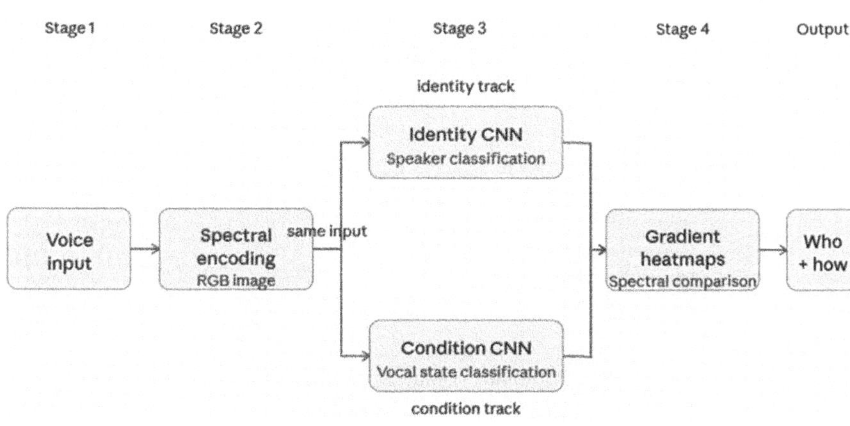

Fig. 1. The model flow

4 Preliminary Results

We performed a preliminary experiment of our model, using a voice samples dataset. The dataset used (RAVDESS – https://zenodo.org/records/1188976), contains recordings from four speakers across five emotional conditions (Amused, Angry, Disgusted, Neutral, Sleepy), yielding 1,382 samples across 1,377 speaker-condition groups. Emotional state labels serve as a proxy for vocal condition more broadly; clinical deployment would require retraining the condition pathway on clinically validated labels relevant to aging, such as fatigue severity or disease progression markers. The identity network is trained on speaker labels; the condition network on emotional state labels. Both tasks are treated as classification problems over the shared spectral image representation. While RAVDESS speakers are not older adults, this evaluation serves as proof-of-feasibility prior to clinical deployment with aging populations.

The identity model achieves 99.64% group-level accuracy (Macro F1: 0.9954). Figures 2 and 3 show the training and validation accuracy and loss curves respectively. Both curves converge cleanly within 20 epochs, with train and chunk validation tracking closely and group validation stabilizing above 0.95 by epoch 10.

In Fig. 2, both train (blue) and chunk (orange) accuracy converge rapidly, both reaching above 0.95 by epoch 7. Group validation accuracy (green) shows greater variance, reflecting the harder generalization task, but stabilizes near 1.0 by epoch 20. The gap between chunk and group validation is expected: group-level evaluation requires the model to generalize across all audio segments of a speaker, not just held-out chunks from seen contexts.

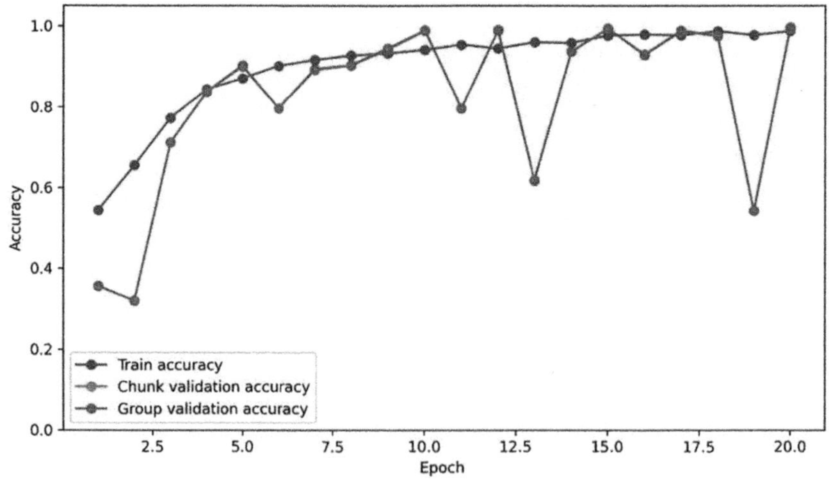

Fig. 2. Identity accuracy over epochs

In Fig. 3, Training loss (blue) decreases monotonically, consistent with effective learning. Chunk validation loss (orange) tracks closely. Group validation loss (green) shows two transient spikes at epochs 13 and 18, suggesting momentary instability in speaker boundary decisions, but recovers fully by epoch 20. Final group validation loss is near zero, consistent with the near-perfect accuracy reported above.

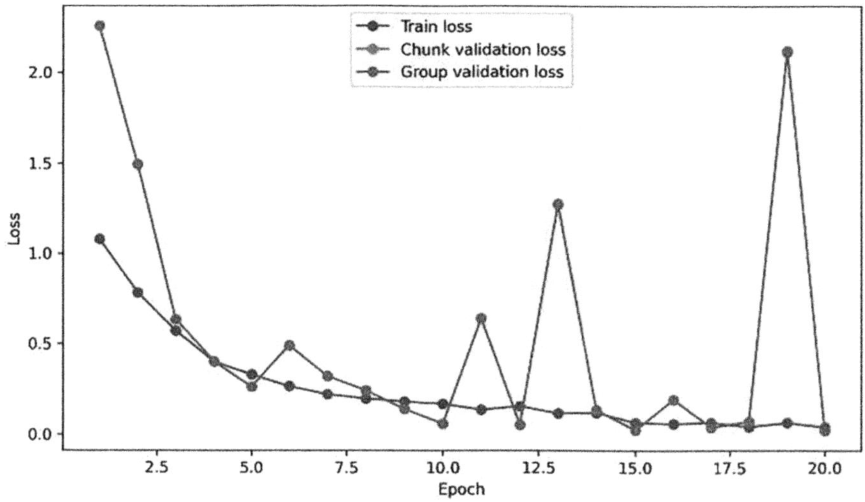

Fig. 3. Identity loss over epochs

Figure 4 presents Identity Group-Level normalized confusion matrix. All four speakers (bea, jenie, josh, sam) are classified with near-perfect accuracy: bea and sam at 1.00, jenie and josh at 0.99. The single off-diagonal value (josh: 0.01 misclassified as sam) represents one group-level error out of the full evaluation set. This result confirms that the shared spectral image representation retains highly discriminative speaker identity information. The near-perfect diagonal in Fig. 4 (identity confusion matrix) confirms that all four speakers are distinguished without misclassification at the group level, demonstrating that audio-to-image encoding preserves sufficient speaker-discriminating information even across varied emotional states (Weighted Precision: 0.99; Weighted Recall: 0.99).

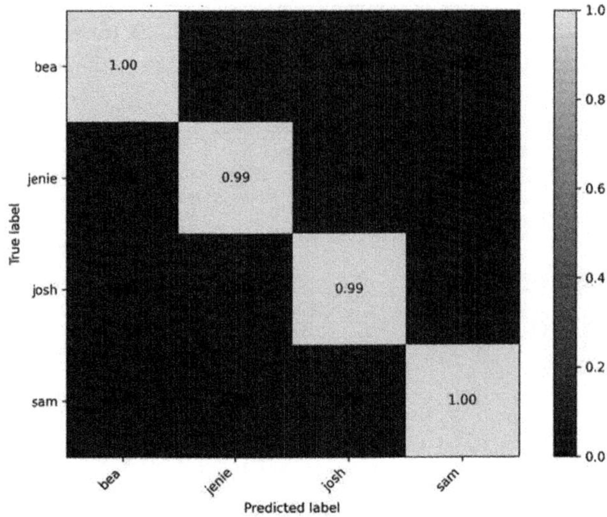

Fig. 4. Identity Group-Level normalized confusion matrix

Next, we examined the condition model. The condition model achieves 94.26% group-level accuracy (Macro F1: 0.9365). Figures 5 and 6 show the corresponding training curves. The condition task is harder than identity: five classes with overlapping acoustic properties (e.g., Amused and Neutral) create genuine ambiguity in the spectral representation. Nevertheless, group validation accuracy reaches above 0.90 by epoch 10 and stabilizes near 0.95 by epoch 20, demonstrating robust learning despite class overlap. Figure 6 (condition confusion matrix) reveals that the model handles Angry (0.96), Neutral (0.99), and Sleepy (0.98) with high precision, while Amused (0.90) and Disgusted (0.84) show modest confusion with each other, which is linguistically and acoustically plausible (Weighted Precision: 0.942, Weighted Recall: 0.942).

Unlike the identity model, the condition model shows a more gradual learning curve, consistent with the higher complexity of the five-class emotional classification task (Fig. 5). Train accuracy (blue) and chunk validation (orange) both rise steadily from around 0.45 at epoch 1 to above 0.90 by epoch 15. Group validation accuracy (green) follows closely, reaching 0.95 by epoch 20 with no sign of overfitting. The smooth convergence suggests the spectral image representation provides stable, learnable features for condition classification.

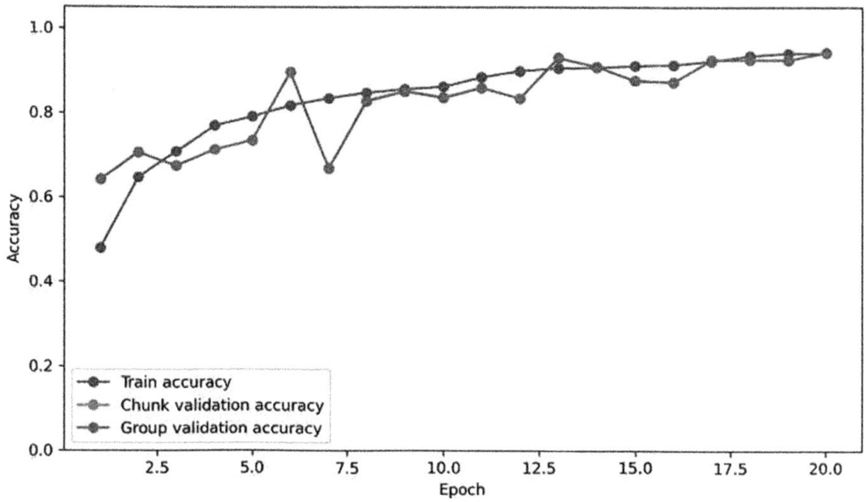

Fig. 5. Condition accuracy over epochs (Color figure online)

Training loss (blue) decreases steadily across all 20 epochs (Fig. 6). A transient spike in group validation loss (green) around epoch 7 indicates a brief instability in generalizing emotional boundaries across speakers, but the model recovers and all three curves converge below 0.25 by epoch 20. The alignment of train and validation loss at convergence indicates the model is not overfitting to training speaker-condition combinations.

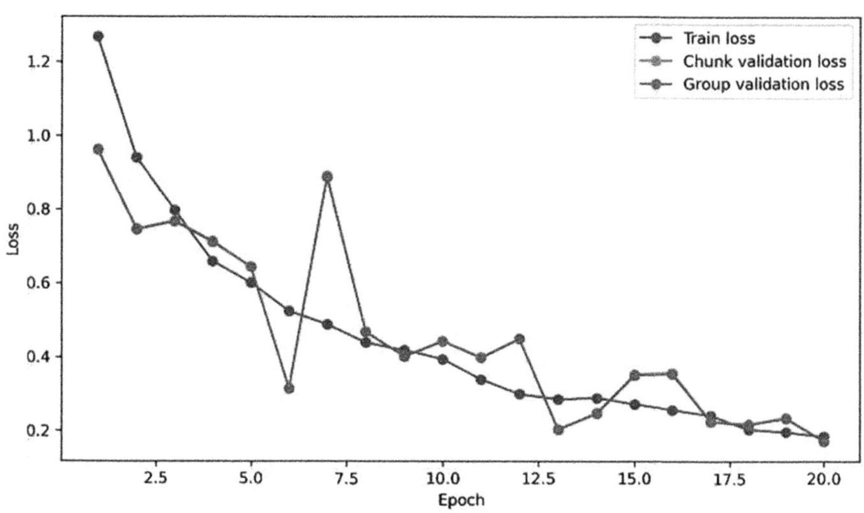

Fig. 6. Condition loss over epochs (Color figure online)

Results are strong across five emotional classes (Fig. 7). Neutral (0.99) and Sleepy (0.98) are classified near perfectly, consistent with their distinct prosodic and spectral signatures. Angry (0.96) and Amused (0.90) perform well, with Amused showing minor confusion with Angry (0.07) and Disgusted (0.02), reflecting genuine acoustic proximity between positively and negatively aroused states. Disgusted achieves 0.84, with small confusions toward Amused (0.06) and Neutral/Sleepy (0.04 each). These patterns are interpretable and consistent with known acoustic overlap in emotional speech literature [15, 21].

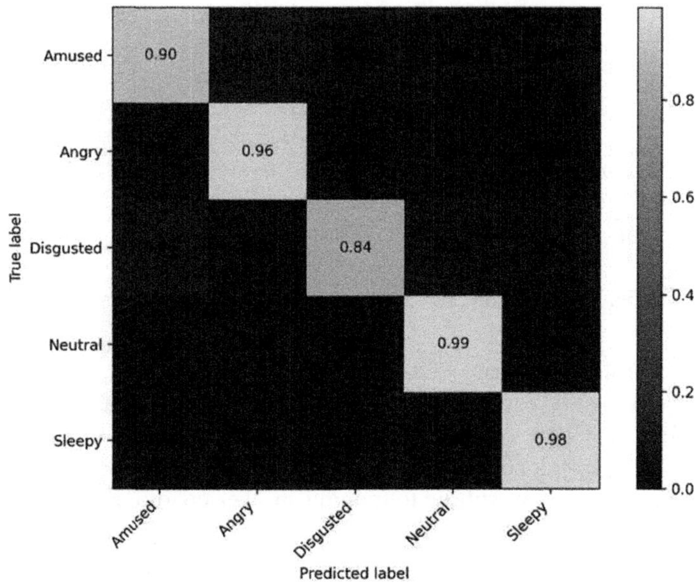

Fig. 7. Condition Group-Level normalized confusion matrix

Taken together, the two models demonstrate that identity and condition classification are jointly tractable from the same spectral image representation, without material degradation of either task. Preliminary visual inspection of attribution maps suggests partial spatial separation across pathways, with condition-related activations appearing to concentrate on lower-frequency prosodic regions and identity-related activations on higher-frequency formant structure. A systematic, quantitative analysis of Grad-CAM heatmaps across all speakers and conditions - including statistical tests of spatial separation—remains as future work and is required before claims about deterministic region-based extraction can be made. Comparison with established baselines such as MFCC-SVM and x-vector models is not included in this preliminary evaluation; the present goal is proof-of-feasibility for the parallel framework rather than competitive benchmarking. Such comparisons, alongside experiments with larger and more diverse speaker populations, are a necessary step before claims of practical effectiveness can be made.

5 What the System Could Hear: Implications for CUI

A system built on this framework would be capable of something current conversational AI cannot do: hear who you are and how you are, simultaneously, from the same utterance. We now turn to what this capability would mean for CUI, with particular attention to aging-related deployment contexts.

5.1 The Identity Assumption Revisited

Conversational interfaces treat the user as a stable, recognizable agent. But aging disrupts this assumption in ways that matter clinically. A person in early-stage dementia is not the same interlocutor on a high-symptom day as on a low-symptom day. A Parkinson's patient whose speech has changed over six months may be misauthenticated by a static voiceprint baseline. An older adult calling a crisis line at 3 a.m. after a fall processes language differently than during a routine interaction [3]. The stable-identity assumption is adequate for most interactions, but in contexts where condition matters, such as eldercare, chronic disease monitoring, fall prevention, and crisis support, the assumption actively prevents the system from being useful. Parallel modeling forces a choice: should conversational systems be designed for the average user in an average state, or for the actual user in their actual state?

5.2 Agency, Adaptation, and Paternalism

If a conversational system can infer that a user is fatigued, distressed, or unwell, what should it do with that inference? Several design responses are possible. The system could adapt its output: simplify language, reduce response length, lower cognitive load. In rehabilitation contexts, this adaptation may be genuinely therapeutic: a system that recognizes vocal markers of fatigue in a post-stroke patient and adjusts its pacing accordingly is responsive rather than paternalistic. It could notify the user of the detected state. It could escalate to professional help, alert a caregiver, or initiate a safety protocol. It could remain silent: detect but not act. Each response encodes a different answer to the same question: whose judgment about the user's state should govern the interaction [12]. For vulnerable older adults, this question has weight: adaptive systems risk undermining autonomy precisely in users for whom autonomy is already under pressure from institutional care.

5.3 Inference from the Body is Different in Kind

Conversational AI routinely infers preferences and intentions from language content. But inferring physiological and emotional state from acoustic properties of voice reaches beneath language to the body. It extracts information the speaker did not choose to communicate. Users who accept that their voice assistant learns their music preferences may feel very differently about a system that detects, stores, and acts on vocal markers of cognitive decline on a Tuesday morning. This is especially salient for older adults in residential care, who may have limited ability to negotiate the terms of surveillance in their living environment.

Nissenbaum's concept of contextual integrity clarifies why this matters [13]: information flows appropriately when they match the norms of the context in which they were shared. A person speaking to a voice assistant to set a reminder has not entered a medical context. Yet a condition-aware system would extract health-relevant inferences from exactly that interaction. Zuboff's analysis [25] offers a sharper framing: vocal condition markers are a particularly intimate form of behavioral surplus, generated not from what users say but from how their bodies sound while saying it.

5.4 Bias, Interpretability, and Aging Voices

The heatmaps produced by the parallel framework are not neutral visualizations. They encode the assumptions of training data as much as they reveal signal structure [12, 15]. This concern is acute for aging populations. Affective computing and paralinguistic corpora are disproportionately drawn from younger adult speakers [15, 21]. A condition classifier trained on such data may systematically misclassify older voices, conflating age-related vocal changes (increased jitter, reduced HNR, slower speech rate) with pathological states. The result would be a system that consistently overcalls distress or disease for healthy older adults, while potentially missing genuine markers in voices that diverge from its training distribution. This is not hypothetical: documented demographic skews in existing corpora make this outcome predictable [15].A heatmap generated by a mis-calibrated model does not reveal where condition information lives in the vocal signal; it reveals where the model learned to look. For older and marginalized speakers, the cost of this limitation is not evenly distributed.

6 Open Questions

On agency: if a system can infer that a user is fatigued, distressed, or unwell, who decides what it does with that inference? A system that simplifies its language when it detects fatigue has already answered that question, quietly, in the design [12]. In eldercare settings, this design choice may be made by institutional providers rather than users themselves, compounding the accountability gap.

On consent: current data governance frameworks apply to data that is explicitly collected. They are not designed for inferences extracted from signals users did not intend as data [13, 26]. Vocal condition markers may be subject to health data protections in some jurisdictions and entirely unprotected in others. Researchers in the field are encouraged to consider how regional regulatory frameworks (GDPR in Europe, HIPAA in the US, emerging AI-specific health regulations) apply to passive physiological inference from voice, and to design systems that are robust to regulatory change rather than dependent on current gaps.

On the limits of identity: the framework targets fatigue, dysphonia, and negative affect. But the boundaries of condition are not fixed. A system that can detect fatigue can, in principle, detect intoxication, grief, or cognitive decline. In aging contexts, the decision about which conditions to model is simultaneously a clinical, ethical, and design question: modeling depression without clinical safeguards, or dementia without caregiver involvement protocols, may cause more harm than the absence of detection.

One further problem cuts across all three. Condition inference does not have a clear failure mode that users can detect. A verification system that fails to recognize a speaker produces a visible error. A condition classifier that misidentifies fatigue as distress produces no such signal. The system acts on the inference, and the user has no mechanism to contest it [12]. For older adults who may already experience reduced agency in care settings, invisible misclassification is not merely an inconvenience: it is a structural harm.

7 Conclusions

Voice is the medium of conversational interfaces. It is also a physiological signal that carries more information than any current conversational system is designed to use. The parallel voice modeling framework we have described is a proposal for hearing more of what is there. For aging populations, hearing more is not an optional enhancement: it is a prerequisite for systems that can support independent living, monitor chronic disease, and respond appropriately to vulnerable users in real time. But hearing more raises questions about agency, consent, and the relationship between a system's inferences and a user's autonomy that the CUI field has not yet fully confronted. The results reported here, 99.64% group-level accuracy for speaker identity (Fig. 4) and 94.26% for vocal condition (Fig. 7), demonstrate that both inference tasks are achievable from the same spectral image representation without mutual degradation. The training dynamics in Figs. 2, 3, 5, and 6 confirm stable convergence in both pathways, while the condition confusion matrix (Fig. 7) reveals interpretable error patterns: the model's residual confusions between Amused, Disgusted, and Angry correspond to genuine acoustic proximity in the emotional speech literature, not noise. These results establish the technical feasibility of the parallel framework under constrained conditions: four speakers, a single healthy-actor corpus, one language, and one recording environment. Scalability to larger and more diverse populations, including clinical cohorts of older adults, remains undemonstrated and constitutes the primary limitation of the current work. The harder question they raise is not whether such a system can be built. It is whether the field is prepared for what building it entails.

We offer a provocation calibrated to this context. Imagine an older adult with a chronic condition speaking to a conversational health assistant at home. The system detects a vocal marker associated with symptom progression. It has three options: tell the user what it has heard, act on the inference without disclosure, or ignore what it has detected because no one decided it should listen. All three options have costs. For aging populations, the costs are not theoretical. The question is not which option is correct. The question is whether the field is ready to choose.

AI Disclosure. The authors used Claude (Anthropic) to assist with drafting and editing portions of this paper. All ideas, arguments, framing, and technical content originated with the authors. AI assistance was used for language refinement and structural editing only. The authors take full responsibility for the content of this paper.

References

1. Atif, Y.: Audio-to-image encoding for improved voice characteristic detection using deep convolutional neural networks. arXiv:2503.05929 (2025)

2. Bouhuys, A.L., Harm, K., Schutte, H., Beersma, D.G.M., Nieboer, G.: Relations between depressed mood and vocal parameters before, during and after sleep deprivation: a circadian rhythm study. J. Affect. Disord. **19**(4), 249–258 (1990)
3. Cummins, N., Scherer, S., Krajewski, J., Schnieder, S., Epps, J., Quatieri, T.F.: A review of depression and suicide risk assessment using speech analysis. Speech Commun. **71**, 10–49 (2015)
4. Dehak, N., Kenny, P.J., Dehak, R., Dumouchel, P., Ouellet, P.: Front-end factor analysis for speaker verification. IEEE Trans. Audio Speech Lang. Process. **19**(4), 788–798 (2010)
5. Dhuheir, M., Albaseer, A., Baccour, E., Erbad, A., Abdallah, M., Hamdi, M.: Emotion recognition for healthcare surveillance systems using neural networks: a survey. In: IWCMC 2021, 681–687 (2021)
6. Fagherazzi, G., Fischer, A., Ismael, M., Despotovic, V.: Voice for health: the use of vocal biomarkers from research to clinical practice. Digit. Biomarkers **5**(1), 78–88 (2021)
7. Garcia-Garcia, M., Caplier, A., Rombaut, M.: Sleep deprivation detection for real-time driver monitoring using deep learning. In: ICIAP 2018, pp. 435–442 (2018)
8. Hecker, P., Steckhan, N., Eyben, F., Schuller, B.W., Arnrich, B.: Voice analysis for neurological disorder recognition: a systematic review. Front. Digit. Health **4**, 842301 (2022)
9. Karthikeyan, V., Priyadharsini, S.S., Balamurugan, K.: Attention-based multi-dimension fused-feature CNN framework for speaker recognition. Multimedia Tools Appl. **84**(31), 38927–38957 (2025)
10. Little, M.A., et al.: Suitability of dysphonia measurements for telemonitoring of Parkinson's disease. IEEE Trans. Biomed. Eng. **56**(4), 1015–1022 (2009)
11. Luxton, D.D., McCann, R.A., Bush, N.E., Mishkind, M.C., Reger, G.M.: MHealth for mental health: integrating smartphone technology in behavioral healthcare. Prof. Psychol. Res. Pract. **42**(6), 505 (2011)
12. Mittelstadt, B.D., Allo, P., Taddeo, M., Wachter, S., Floridi, L.: The ethics of algorithms: mapping the debate. Big Data Soc. **3**(2), 2053951716679679 (2016)
13. Nissenbaum, H.: Privacy as contextual integrity. Wash. L. Rev. **79**, 119 (2004)
14. Reynolds, D.A., Quatieri, T.F., Dunn, R.B.: Speaker verification using adapted Gaussian mixture models. Digit. Signal Process. **10**(1–3), 19–41 (2000)
15. Schuller, B., et al.: The INTERSPEECH 2013 computational paralinguistics challenge: social signals, conflict, emotion, autism. In: INTERSPEECH 2013, 148–152 (2013)
16. Selvaraju, R.R., Cogswell, M., Das, A., Vedantam, R., Parikh, D., Batra, D.: Grad-CAM: visual explanations from deep networks via gradient-based localization. In: Proceedings of the IEEE International Conference on Computer Vision, pp. 618–626 (2017)
17. Snyder, D., Garcia-Romero, D., Sell, G., Povey, D., Khudanpur, S.: X-vectors: robust DNN embeddings for speaker recognition. In: 2018 IEEE International Conference on Acoustics, Speech and Signal Processing (ICASSP), pp. 5329–5333. IEEE (2018)
18. Tandel, N.H., Prajapati, H.B., Dabhi, V.K.: Voice recognition and voice comparison using machine learning techniques: a survey. In: 2020 6th International Conference on Advanced Computing and Communication Systems (ICACCS), pp. 459–465. IEEE (2020)
19. Thoret, E., Andrillon, T., Gauriau, C., Léger, D., Pressnitzer, D.: Sleep deprivation measured by voice analysis. bioRxiv (2022)
20. Tsanas, A., Little, M., McSharry, P., Ramig, L.: Accurate telemonitoring of Parkinson's disease progression by non-invasive speech tests. Nat. Proc. 1 (2009)
21. Tzirakis, P., Trigeorgis, G., Nicolaou, M.A., Schuller, B.W., Zafeiriou, S.: End-to-end multimodal emotion recognition using deep neural networks. IEEE J. Sel. Top. Signal Process. **11**(8), 1301–1309 (2017)
22. World Health Organization. Ageing and health. WHO Fact Sheet(2023)
23. Wright, H., Aharonson, V.: Vocal feature changes for monitoring Parkinson's disease progression. Brain Sci. **15**(3), 320 (2025)

24. Zacharia, T., D'souza, P., Mathew, M., D'souza, G., James, J., Baliga, M.S.: Effect of circadian cycle on voice: a cross-sectional study with young adults of different chronotypes. J. Laryngol. Voice **8**(1), 19–23 (2018)
25. Zuboff, S.: The age of surveillance capitalism. In: Social Theory Rewired, pp. 203–213. Routledge (2023)

Age and Skin Tone Biases in Remote Heart Rate Monitoring

Boyi Ma[1,2(✉)], Jathushan Kaetheeswaran[1,2,3], Ali Abedi[4], Milad Lankarany[1,2,3], and Shehroz S. Khan[5(✉)]

[1] Institute of Biomedical Engineering, University of Toronto, Toronto, ON, Canada
boyima.ma@mail.utoronto.ca
[2] KITE Research Institute, Toronto Rehabilitation Institute, University Health Network, Toronto, ON, Canada
[3] Krembil Brain Institute, University Health Network, Toronto, ON, Canada
[4] Peter Munk Cardiac Centre, University Health Network, Toronto, ON, Canada
[5] College of Engineering and Technology, American University of the Middle East, Egaila 54200, Kuwait
Shehroz.Khan@aum.edu.kw

Abstract. Remote photoplethysmography (rPPG) enables contactless heart rate monitoring from facial video, offering significant promise for accessible heart rate monitoring. Despite rapid progress in deep learning-based rPPG, the effect of demographic factors on model performance needs to be further explored. Existing studies evaluate a few models across inconsistent datasets and focus mainly on skin tone, leaving age bias largely uncharacterized. In this work, we conduct a demographic bias evaluation of five representative rPPG deep learning models on HEART-Video, an inclusive dataset with balanced demographic coverage. We investigate age and skin tone demographic factors, using a consistent cross-group evaluation protocol across three physical activities. Our results reveal that all five models exhibit consistent demographic biases regardless of architecture. All the models trained on younger participants fail to generalize to older populations. Skin tone bias produces significant performance gaps and is amplified under motion-heavy conditions. Our findings establish the empirical foundation necessary for designing targeted, fairness-aware rPPG systems and demonstrate that equitable contactless cardiovascular monitoring for aging populations demands age-aware training, evaluation, and deployment protocols.

Keywords: Remote photoplethysmography · Demographic bias · Heart rate estimation · Fairness evaluation · Equitable AI

1 Introduction

Cardiovascular disease (CVD) remains the leading cause of death worldwide [1]. Early prevention and timely intervention of CVD is increasingly critical across

S. S. Khan et al. (Eds.): IJCAI 2026, CCIS 3127, pp. 138–151, 2027.
https://doi.org/10.1007/978-981-92-5190-2_10

all age groups, including children and young adults [2]. Heart rate (HR) is a fundamental physiological signal and a key indicator of cardiovascular condition, making continuous HR monitoring a valuable clinical tool for cardiovascular health [3]. Electrocardiography (ECG) is the clinical gold standard for HR measurement. However, it requires specialized equipment and clinical supervision, which limits accessibility and mobility [4]. Photoplethysmography (PPG) offers a non-invasive alternative by measuring blood volume changes through light absorption on the surface of skin [5]. While PPG sensors can be worn on the finger, wrist, or ear, contact-based measurement can cause discomfort or skin irritation and is unsuitable for individuals with cognitive or sensory sensitivities. Remote PPG (rPPG) addresses these limitations by providing a contact-free solution that extracts HR from subtle color changes in facial video. With the rapid advancement of artificial intelligence, rPPG methods have advanced rapidly, progressing from traditional signal processing approaches to data-driven deep learning architectures.

Despite significant progress, the impact of demographic factors on rPPG performance remains underexplored. Prior studies suggest that demographic attributes can significantly influence performance. Cao et al. [6] show that age significantly influences blood oxygen saturation estimation using rPPG data. Nowara et al. [7] analyze four traditional and one deep learning model, finding a notable accuracy drop for the darkest skin tones. However, existing studies do not characterize how and to what extent demographic factors affect the performance of state-of-the-art rPPG deep learning models. The nature and scale of demographic bias must be understood before bias mitigation strategies can be designed. Without the evaluation, the field risks deploying rPPG systems that perform inequitably across populations, undermining the promise of accessible, contactless cardiovascular monitoring. To address this, we conduct a comprehensive demographic bias evaluation of five representative rPPG deep learning models on a novel inclusive rPPG dataset. We examine age and skin tone factors using a consistent experimental protocol. Our goal is to quantify performance disparities and provide insights into fairness in rPPG systems. Understanding these biases is a necessary step toward building equitable and reliable AI-driven healthcare solutions.

Our contributions are as follows:

- We establish an empirical benchmark on an inclusive rPPG dataset to reveal the presence of demographic biases, with a particular focus on age and skin tone disparities in remote HR estimation.
- We conduct thorough evaluations using five state-of-the-art deep learning-based rPPG models for HR estimation, demonstrating that age and skin tone biases consistently persist across different model architectures.

2 Related Work

Recent years have seen rapid progress in HR estimation using deep learning-based rPPG methods. Early work such as PhysNet [8] introduced deep spatio-temporal convolutions networks to rPPG and remains a widely used baseline. PhysFormer [9] extended this to a transformer architecture-based design, using temporal difference modeling to capture both global and local representations, establishing a representative architecture in the field. RhythmFormer [10] further improves attention modeling with conventional and periodic sparse attention modules. Beyond these, recent work has incorporated gated recurrent units [11] and 3D facial maps from improved region-of-interest selection [12]. Despite this architectural diversity, they largely overlook fairness and the effect of demographic factors on model performance, which leaves a gap that directly limits the equity of deployed rPPG systems.

Public datasets play a central role in advancing rPPG research. However, recent analyses reveal significant limitations in demographic diversity. Khan and Mostafa [13] review 32 rPPG datasets and find that most lack balanced demographic data across age, sex, and skin tone. Similarly, Bondarenko et al. [14] quantitatively analyzes 100 rPPG studies and demonstrates significant underrepresentation of darker skin tones and gender imbalance across the field. These limitations potentially hinder the study of demographic bias. A dataset with comprehensive and balanced demographic coverage is essential for a thorough demographic bias investigation. Table 1 compares several datasets that release demographic information. Among them, HEART-Video provides the broadest range of ages with diverse ethnicities, making it well-suited for this study.

A small number of studies have begun to investigate and address demographic bias in rPPG. Dasari et al. [25] collects data from darker-skinned participants and evaluates four signal processing methods alongside two early deep learning models, and report performance disparities. Ba et al. [26] addresses skin tone bias through synthetic augmentation, translating light-skinned participants to darker tones using style transfer. Comas et al. [27] incorporates CIELAB skin color features into a deep learning model to reduce HR errors for darker-skinned individuals [28].

While these efforts represent progress, they share a common limitation: bias mitigation is pursued without a clear understanding of the bias characterization. Moreover, majority of the bias studies focus on skin tone, leaving the effects of age largely unexplored. A thorough evaluation across multiple demographic factors on state-of-the-art models is needed before equitable rPPG systems can be built.

3 Methodology

The task considered in this study is supervised HR estimation from facial videos. Specifically, given an input facial video sequence, the rPPG model predicts the corresponding HR value. Since heart rate is a continuous variable, the task is formulated as a regression problem. To investigate the influence of demographic factors on HR estimation, we design a controlled experimental framework that isolates the effects of age and skin tone on model performance.

Table 1. Comparison of several publicly available rPPG datasets with demographic information.

Dataset	Number of Videos (duration)	Ground Truth	Age	Sex	Skin Tone/Ethnicity
MAHNOB [15]	540 (10 s)	Heart Rate	19–40 years (26.06 ± 4.39)	13M, 16F	Different cultural backgrounds
COHFACE [16]	160 (1 min)	PPG signal	15 – 65 years (25.6 ± 11.5)	28M, 12F	–
MMSE-HR [17]	102 (20 s)	Heart Rate	18 – 66 years	58M, 82F	Black, White, Asian, Hispanic, and Others
VIPL-HR-V2 [18]	2500 (30 s)	Heart Rate	35.4 ± 18.0	49%M, 51%F	Asians
VicarPPG-2 [19]	40 (5 min)	Heart Rate	20–35 years (29 ± 5)	7M, 3F	Fitzpatrick scale from 2 to 4
iBVP [20]	128 (3 min)	PPG signal	18–45 years (27 ± 5.8)	10M, 23F	Asian, African, White, Mixed, Others
V4V [21]	1358 (1 min)	Heart Rate	18–66 years	101M, 78F	African American, White, Asian, Hispanic, and Native American
VITAL [22]	432 (2 min)	Heart Rate	26–41 years	33M, 21F	White, Asian, Black, American Indian, and Unknown
CHILL [23]	180 (1 min)	PPG signal	18–32 years	17M, 28F	Fitzpatrick scale of 1–3
HEART-Video [24]	960 (30 s)	Heart Rate	19–75 years	17M, 23F	White, South Asian, East Asian, Middle Eastern, Latino, Black

3.1 Dataset Description

We conduct our study on HEART-Video, a novel and demographically diverse rPPG dataset [24]. The dataset contains facial videos from 40 healthy participants, along with synchronized ECG signals as ground truth. It includes 17 males and 23 females ranging from 19 to 75 years old, spanning White, Asian, Latino, Middle Eastern, and Black ethnicities, providing diversity in skin tones. Each participant is recorded under three physical states—sitting, standing, and walking—for 4 min each, resulting in a total of 480 min of video. All recordings are segmented into 30-s clips, producing 960 video segments in total. Ground truth HR is extracted from the ECG signals. The setup of the HEART-Video dataset is illustrated in Fig. 1.

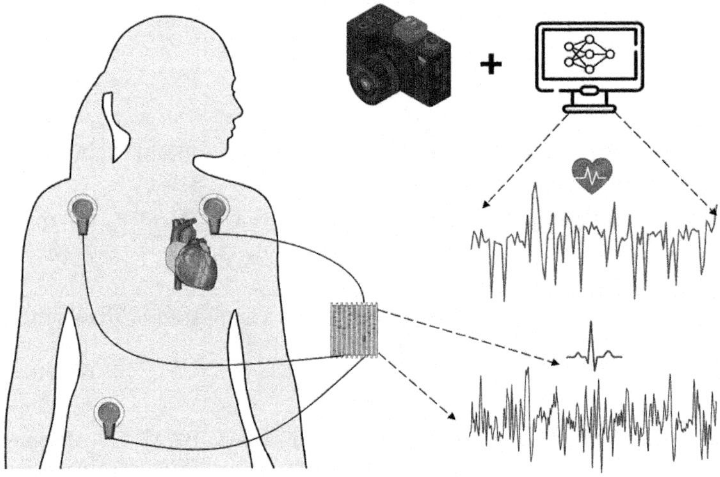

Fig. 1. Visualization of HEART-Video Dataset Setup.

3.2 Deep Learning Models

We evaluate five representative rPPG models spanning from 2019 to 2025. The selection covers CNN-based and transformer-based architectures and includes both widely used baselines and recent state-of-the-art methods. This diverse selection allows us to analyze whether demographic bias persists across different model designs. Brief descriptions of each model are provided below.

- PhysNet [8] is the first rPPG method to adopt deep spatio-temporal convolutional networks for end-to-end rPPG signal reconstruction from raw facial videos.
- PhysFormer [9] introduces a transformer-based architecture that leverages temporal difference transformers to capture both local and global spatio-temporal features.

- EfficientPhys [29] focuses on lightweight and efficient physiological measurement that eliminates preprocessing steps such as face detection, segmentation, and color space transformation.
- FactorizePhys [30] proposes a self-attention module that jointly models spatial, temporal, and channel-wise dependencies through nonnegative matrix factorization.
- MMRPhys [31] is a multimodal 3D-CNN architecture for simultaneous multitask estimation of rPPG and respiratory signals from RGB and thermal videos, with the flexibility to operate using one single modality alone.

3.3 Experiments Design

- *Age Bias* – Following the United Nations guidelines [32], we use 60 years as the threshold to divide participants into young (24 participants) and old (16 participants) groups. Each group is further randomly split into two folds, ensuring all segments from one participant remain in the same fold. We train each model on young_fold1 and test on both young_fold2 and old_fold, then repeat with young_fold2 as the training set. The same protocol is applied with the older group.
- *Skin Tone Bias* – Participant ethnicity is self-reported and does not directly provide skin tone labels. Manual labeling skin tones can introduce subjective bias and limits reproducibility. To address this issue, we follow the universal evaluation benchmark [33] and apply an automated skin tone classifier [34] to assign consistent labels to all participants. The resulting skin tone distribution is shown in Fig. 2. Skin tone labels depend on recording conditions such as lighting and illumination, making it difficult to compare across datasets collected using different protocols. However, these factors internally consistent within the HEART-Video dataset. Based on this classification, we divide participants into a light skin tone group (28 participants) and a dark skin tone group (12 participants) and apply the same fold-splitting and cross-group testing protocol used for age.

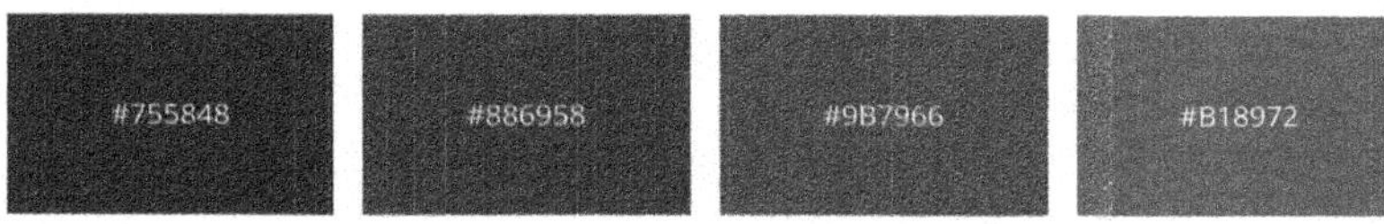

Fig. 2. Visualization of Skin Tone Coverage and Examples in HEART-Video.

3.4 Implementation Details

All videos are processed at the original frame rate of 30 fps. Face regions are detected and cropped using Google MediaPipe [35], a real-time face landmark detection pipeline that identifies 468 facial landmarks and enables precise region-of-interest localization. Because the walking condition introduces significant head movement, we perform face detection every 30 frames (i.e., per second) instead of once per video, ensuring accurate spatial alignment throughout. Cropped frames are further normalized using standard z-score normalization and resized to 72×72 pixels. As predicting HR is a regression task and following the protocol from the rPPG-Toolbox [36], each prediction corresponds to a 6-second window of 180 frames. All models are trained for 30 epochs using the Adam optimizer with a learning rate of 1×10^{-3} and mean squared error loss. We report performance using Mean Absolute Error (MAE), which measures the average magnitude of HR prediction errors in beats per minute (bpm), providing an absolute measure of deviation. We also conducted statistical analyses across all experimental settings. For each sensitive attribute (e.g., age and skin tone), the MAE obtained from each model under different activities was treated as a paired observation. A paired t-test was performed to compare the prediction errors between demographic groups. Statistical significance was determined at $p < 0.05$.

4 Experiments

4.1 Age Bias

Tables 2 report MAE for models trained on young participants and evaluated on both young and old test sets across three activities. The t-test results show that the prediction error for older participants was significantly higher than that for younger participants (t = 6.16, $p < 0.001$). A consistent cross-age performance drop is observed across all models and metrics. When trained on young_fold1 and tested on the old group, MAE increases substantially across all models and activities, confirming that models trained exclusively on young participants fail to generalize to older populations. This degradation persists regardless of which young fold is used for training, indicating the gap is systematic rather than fold-specific. PhysNet achieves the lowest in-group errors under sitting and standing conditions, yet exhibits notable degradation on old participants, suggesting that strong in-distribution performance does not imply equitable cross-age transfer. PhysFormer demonstrates comparatively stable cross-age behavior under sitting, though this advantage diminishes during standing and walking. MMRPhys displays the most inconsistent pattern under different conditions, indicating high sensitivity to activity context.

Tables 3 show that models trained on older participants presented improved performance on younger participants in several cases. The t-test analysis confirmed the presence of performance differences across the age groups (t = 4.25, $p < 0.001$). We attribute this to two factors. First, older adults exhibit higher intra-group physiological variability. Training on a lower-variance distribution on young participants fails to generalize to a high-variance distribution for older adults. Second, aging-related facial texture varies, including wrinkles and uneven pigmentation, directly degrade signal quality for models. Therefore, systems trained predominantly on younger populations, as is typical in current rPPG benchmarks, underperform for older adults.

Table 2. MAE results of five rPPG models trained on young_fold1 (young_fold2) and evaluated on young_fold2 (young_fold1) and old group across three activity conditions.

Model	Activity	Young_fold1		Young_fold2 (train set)	
		Young_fold2	Old	Young_fold1	Old (test set)
PhysNet	Sitting	10.83 ± 0.25	16.37 ± 0.37	7.75 ± 0.34	16.20 ± 0.33
PhysFormer		12.61 ± 0.30	14.21 ± 0.26	10.75 ± 0.44	14.76 ± 0.30
EfficientPhys		11.10 ± 0.38	12.52 ± 0.22	11.06 ± 0.36	13.08 ± 0.22
FactorizePhys		12.12 ± 0.46	13.44 ± 0.39	17.58 ± 1.23	19.19 ± 0.76
MMRPhys		22.16 ± 0.80	33.09 ± 1.12	9.12 ± 0.35	20.23 ± 0.66
PhysNet	Standing	10.31 ± 0.34	20.38 ± 0.47	9.73 ± 0.41	20.98 ± 0.51
PhysFormer		10.75 ± 0.34	17.87 ± 0.36	10.80 ± 0.33	19.48 ± 0.43
EfficientPhys		11.38 ± 0.31	17.07 ± 0.32	10.50 ± 0.33	19.38 ± 0.34
FactorizePhys		12.79 ± 0.45	19.56 ± 0.50	18.42 ± 1.01	18.58 ± 0.60
MMRPhys		10.56 ± 0.27	16.11 ± 0.42	23.53 ± 1.03	17.23 ± 0.57
PhysNet	Walking	10.29 ± 0.34	17.73 ± 0.52	11.70 ± 0.26	15.57 ± 0.52
PhysFormer		9.34 ± 0.31	16.19 ± 0.51	13.43 ± 0.37	14.85 ± 0.40
EfficientPhys		10.41 ± 0.43	13.52 ± 0.36	12.47 ± 0.26	14.38 ± 0.36
FactorizePhys		11.36 ± 0.38	13.37 ± 0.42	13.12 ± 0.34	14.28 ± 0.45
MMRPhys		14.03 ± 0.53	15.41 ± 0.48	21.62 ± 0.50	23.30 ± 1.14

Table 3. MAE results of five rPPG models trained on old_fold1 (old_fold2) and evaluated on old_fold2 (old_fold1) and the young group across three activity conditions.

Model	Activity	Old_fold1		Old_fold2 (train set)	
		Old_fold2	Young	Old_fold1	Young (test set)
PhysNet	Sitting	11.03 ± 0.60	13.37 ± 0.30	6.57 ± 0.46	12.46 ± 0.34
PhysFormer		15.40 ± 0.72	17.95 ± 0.38	10.38 ± 0.43	13.18 ± 0.34
EfficientPhys		10.55 ± 0.34	11.19 ± 0.26	8.39 ± 0.37	14.10 ± 0.32
FactorizePhys		10.67 ± 0.77	17.14 ± 0.69	19.19 ± 1.43	17.84 ± 0.70
MMRPhys		19.58 ± 0.90	13.00 ± 0.29	11.26 ± 0.60	16.52 ± 0.40
PhysNet	Standing	15.09 ± 0.51	18.86 ± 0.44	9.84 ± 0.28	15.77 ± 0.31
PhysFormer		17.73 ± 0.87	23.27 ± 0.43	10.92 ± 0.42	19.97 ± 0.39
EfficientPhys		13.79 ± 0.52	12.40 ± 0.27	10.10 ± 0.51	17.40 ± 0.38
FactorizePhys		20.81 ± 0.94	26.21 ± 0.63	27.63 ± 1.28	26.28 ± 0.88
MMRPhys		42.70 ± 1.23	44.88 ± 0.94	23.37 ± 1.28	32.04 ± 0.96
PhysNet	Walking	14.96 ± 0.55	14.60 ± 0.37	10.76 ± 0.32	14.57 ± 0.31
PhysFormer		16.88 ± 0.70	16.52 ± 0.43	7.49 ± 0.24	14.73 ± 0.36
EfficientPhys		13.15 ± 0.68	11.50 ± 0.27	10.92 ± 0.25	12.02 ± 0.31
FactorizePhys		13.99 ± 0.77	13.74 ± 0.37	9.32 ± 0.39	13.88 ± 0.36
MMRPhys		18.54 ± 1.08	18.15 ± 0.44	20.19 ± 0.92	22.84 ± 0.71

4.2 Skin Tone Bias

Tables 4 report MAE for models trained on light skin participants and evaluated on both light and dark test sets across sitting, standing, and walking. Training on light skin participants and evaluating on dark skin participants produces the largest and most consistent performance gap observed across all demographic bias experiments. The t-test results indicate that the MAE for dark skin participants was significantly higher than that for light skin participants ($t = 7.49$, $p < 0.001$). Under sitting, PhysNet MAE increases from 14.06 (trained on light_fold1 and tested on light_fold2) to 16.90 (tested on dark group), while FactorizePhys rises from 15.43 to 20.29. The degradation intensifies under standing: MMRPhys MAE climbs from 14.36 to 31.31 (increases 118.04%), and FactorizePhys from 15.60 to 23.90. These gaps establish skin tone as the strongest source of demographic bias in this evaluation. Standing produces the most severe degradation across all models, with MMRPhys and FactorizePhys showing catastrophic error inflation. Walking narrows the relative gap for PhysNet and EfficientPhys but amplifies it for MMRPhys and FactorizePhys. Sitting shows intermediate behavior. This non-uniform activity dependence suggests that skin tone bias interacts differently with motion artifacts depending on model architecture.

Table 4. MAE results of five rPPG models trained on light_fold1 (light_fold2) and evaluated on light_fold2 (light_fold1) and the dark group across three activity conditions.

Model	Activity	Light_fold1		Light_fold2 (train set)	
		Light_fold2	Dark	Light_fold1	Dark (test set)
PhysNet	Sitting	14.06 ± 0.38	16.90 ± 0.48	10.18 ± 0.23	14.69 ± 0.41
PhysFormer		12.48 ± 0.30	13.67 ± 0.43	11.50 ± 0.28	15.46 ± 0.52
EfficientPhys		11.57 ± 0.33	12.04 ± 0.41	9.41 ± 0.23	14.40 ± 0.46
FactorizePhys		15.43 ± 0.59	20.29 ± 0.86	18.13 ± 0.75	20.13 ± 0.92
MMRPhys		12.89 ± 0.31	18.71 ± 0.68	14.04 ± 0.33	17.23 ± 0.63
PhysNet	Standing	15.32 ± 0.39	24.87 ± 0.56	11.13 ± 0.27	21.23 ± 0.58
PhysFormer		13.55 ± 0.38	22.92 ± 0.63	11.55 ± 0.30	18.25 ± 0.50
EfficientPhys		12.00 ± 0.38	16.15 ± 0.46	11.53 ± 0.27	17.20 ± 0.48
FactorizePhys		15.60 ± 0.56	23.90 ± 0.84	13.35 ± 0.46	19.48 ± 0.68
MMRPhys		14.36 ± 0.37	31.31 ± 1.31	9.71 ± 0.36	26.79 ± 0.68
PhysNet	Walking	12.99 ± 0.39	13.75 ± 0.52	9.63 ± 0.30	17.54 ± 0.49
PhysFormer		12.75 ± 0.33	15.56 ± 0.58	8.93 ± 0.33	15.46 ± 0.53
EfficientPhys		12.65 ± 0.43	13.11 ± 0.50	10.23 ± 0.27	13.03 ± 0.50
FactorizePhys		11.39 ± 0.36	14.64 ± 0.62	9.48 ± 0.31	14.68 ± 0.60
MMRPhys		14.58 ± 0.47	17.86 ± 0.84	13.19 ± 0.47	19.49 ± 0.64

Table 5. MAE results of five rPPG models trained on dark_fold1 (dark_fold2) and evaluated on dark_fold2 (dark_fold1) and the light group across three activity conditions.

Model	Activity	Dark_fold1		Dark_fold2 (train set)	
		Dark_fold2	Light	Dark_fold1	Light (test set)
PhysNet	Sitting	9.75 ± 0.43	14.81 ± 0.28	9.48 ± 0.58	33.12 ± 0.69
PhysFormer		8.29 ± 0.46	10.76 ± 0.22	22.49 ± 0.68	10.33 ± 0.26
EfficientPhys		8.62 ± 0.39	12.41 ± 0.21	11.67 ± 0.39	15.01 ± 0.24
FactorizePhys		16.85 ± 1.64	19.64 ± 0.65	11.76 ± 0.81	24.60 ± 0.63
MMRPhys		18.41 ± 1.00	19.91 ± 0.47	41.15 ± 1.46	32.98 ± 0.78
PhysNet	Standing	11.09 ± 0.62	17.27 ± 0.41	18.21 ± 0.95	22.69 ± 0.42
PhysFormer		15.43 ± 0.54	11.50 ± 0.27	29.54 ± 0.79	12.85 ± 0.31
EfficientPhys		11.30 ± 0.62	15.51 ± 0.25	11.87 ± 0.42	17.36 ± 0.29
FactorizePhys		29.71 ± 2.36	24.19 ± 0.69	27.28 ± 1.62	24.58 ± 0.66
MMRPhys		12.84 ± 0.47	20.69 ± 0.44	18.94 ± 0.77	13.90 ± 0.31
PhysNet	Walking	11.70 ± 0.40	13.94 ± 0.24	18.28 ± 1.06	15.68 ± 0.36
PhysFormer		11.11 ± 0.62	12.99 ± 0.37	40.14 ± 1.07	19.69 ± 0.43
EfficientPhys		12.45 ± 0.46	14.64 ± 0.23	11.10 ± 0.78	13.94 ± 0.22
FactorizePhys		10.17 ± 0.40	12.48 ± 0.26	13.22 ± 0.82	12.89 ± 0.24
MMRPhys		19.81 ± 0.84	17.37 ± 0.55	37.53 ± 1.83	26.26 ± 0.59

Tables 5 report results for models trained on dark skin participants. Even though no statistically significant difference was observed between dark skin and light skin participants (t = 0.08, p = 0.936). The dark-group training produces the most extreme errors. PhysNet MAE reaches 33.12 from 9.48 (trained on dark_fold2 and tested on light group) under sitting, which is more than three times its in-distribution performance. Light absorption and scattering properties of skin tissue vary substantially across populations, directly affecting the amplitude and spectral composition of rPPG signals. Models trained exclusively on light skin participants learn signal representations calibrated to higher signal-to-noise ratios characteristic, rendering them less accurate on darker skin tones. The severity and consistency of the observed bias confirm that skin tone bias is a real and substantial challenge in HR estimation.

5 Discussion

Demographic biases are real and measurable, while they differ in magnitude and mechanism. Age bias is driven by physiological signal changes across the lifespan and skin tone bias is rooted in optical physics. Each factor introduces distinct challenges that affect signal quality and model generalization in different ways. Consequently, equitable rPPG systems will require targeted mitigation strategies specifically designed for the underlying source of bias. The present study establishes the empirical foundation necessary for developing and prioritizing such fairness-aware interventions, particularly for reliable deployment in aging and real-world healthcare settings.

This interaction between motion and age with large predictions errors is clinically critical for older adults who engage in mobility-related activities, which can have implications in fall risk assessment and rehabilitation monitoring. The existence of skin tone bias in detecting HR highlights the inherent inequity in building AI models. The results show that the current deep learning architectures do not appear to address the compounding effect of motion, age and skin tone in their implementations; thus their real-world deployment capabilities remain unknown.

This study has several limitations. First, the training and testing sets have imbalanced demographic distributions; the observed degradation may result from both demographic bias and imbalanced representation during training and evaluation. Future work should incorporate rigorously balanced training and testing sets for a clearer analysis of these factors. Second, we do not explicitly investigate whether skin tone and aging-related facial texture, such as wrinkles and speckles, contribute to the reduced age-related performance degradation. These appearance variations may affect light reflection and pulse signal extraction. Future work could analyze the relationship between facial texture characteristics and rPPG signal quality. Third, skin tone labels are generated using an automated skin tone classifier for consistency and reproducibility. However, the classifier itself may introduce bias which may affect the precision of the skin tone analysis. Finally, all the experiments are conducted under one single acquisition setup and focus only on HR estimation accuracy. The fairness behavior of rPPG models under different devices, illumination conditions, and waveform-level evaluations remains unexplored. Future work should investigate cross-environment robustness and develop fairness-aware mitigation strategies for equitable rPPG systems.

6 Conclusion

We present a thorough evaluation of demographic bias in rPPG across age and skin tone. Using a consistent cross-group protocol on HEART-Video dataset, we evaluate five representative deep learning models under sitting, standing, and walking conditions. Through controlled experiments across age and skin tone, we demonstrated that demographic bias is a consistent and measurable problem in all five state-of-the-art deep learning models. Our results demonstrate consistent and significant performance degradation on older adults when training solely on young populations. Skin tone bias also produces severe gaps, which states that models trained on lighter skin tones have not learned the representations that can accommodate darker skin tones. We further observed that activity conditions and model architecture influence bias magnitude, particularly under motion-heavy scenarios such as standing and walking.

From an aging and healthcare perspective, these findings carry important implications. Older adults are among the primary target populations for continuous physiological monitoring in rehabilitation, assisted living, and long-term care settings. However, our results show that existing rPPG systems are least reliable for these populations. Without explicit demographic evaluation, AI-based physiological sensing systems risk amplifying healthcare inequities in real-world deployment. There are several directions for future work, including analyzing the relationship between facial texture characteristics and rPPG signal quality, as well as assessing the influence of potentially confounding demographic attributes using a rigorously balanced dataset. The fairness behavior of rPPG models under different devices, illumination conditions, and waveform-level evaluations remains unexplored. Future work should investigate cross-environment robustness and develop fairness-aware mitigation strategies for equitable rPPG systems. From a predictive modeling perspective, bias-aware and fairness-based AI models need to be developed that can mitigate the role of age, skin tone and body movements in estimating HR from facial videos. Overall, this work establishes an empirical benchmark for demographic fairness evaluation in rPPG and highlights the necessity of age-aware and fairness-oriented model design. We believe that these findings motivate future research toward robust, equitable, and clinically reliable contactless physiological monitoring systems for aging populations.

References

1. Guo, J.W., et al.: Cardiovascular health changes in young adults and risk of later-life cardiovascular disease. JAMA Netw. Open **8**(10), 2535573 (2025)
2. Perret-Guillaume, C., Joly, L., Benetos, A.: Heart rate as a risk factor for cardiovascular disease. Prog. Cardiovasc. Dis. **52**(1), 6–10 (2009)
3. Galli, A., Montree, R.J., Que, S., Peri, E., Vullings, R.: An overview of the sensors for heart rate monitoring used in extramural applications. Sensors **22**(11), 4035 (2022)
4. Mastoi, Q.-U.-A., et al.: Heart patient health monitoring system using invasive and non-invasive measurement. Sci. Rep. **14**(1), 9614 (2024)
5. Park, J., Seok, H.S., Kim, S.-S., Shin, H.: Photoplethysmogram analysis and applications: an integrative review. Front. Physiol. **12**, 808451 (2022)
6. Cao, Y., Li, Y., Yiu, K.H., Lam, E.Y.: Addressing healthcare needs of the elderly in contactless blood oxygen saturation estimation with rppg. Comput. Optical Imaging Artif. Intell. Biomed. Sci. **12857**, 55–59 (2024)
7. Nowara, E.M., McDuff, D., Veeraraghavan, A.: A meta-analysis of the impact of skin tone and gender on non-contact photoplethysmography measurements. In: Proceedings of the IEEE/CVF Conference on Computer Vision and Pattern Recognition Workshops, pp. 284–285 (2020)
8. Yu, Z., Li, X., Zhao, G.: Remote photoplethysmograph signal measurement from facial videos using spatio-temporal networks. In: Proceedings of BMVC (2019)
9. Yu, Z., Shen, Y., Shi, J., Zhao, H., Torr, P.H., Zhao, G.: Physformer: facial video-based physiological measurement with temporal difference transformer. In: Proceedings of the IEEE/CVF Conference on Computer Vision and Pattern Recognition, pp. 4186–4196 (2022)

10. Zou, B., Guo, Z., Chen, J., Zhuo, J., Huang, W., Ma, H.: Rhythmformer: extracting patterned rppg signals based on periodic sparse attention. Pattern Recogn. **164**, 111511 (2025)
11. Yao, C., Ren, J., Bai, R., Du, H., Liu, J., Jiang, X.: Mask attack detection using vascular-weighted motion-robust rppg signals. IEEE Trans. Inf. Forensics Secur. **18**, 4313–4328 (2023)
12. Wong, K.L., et al.: Optimising rppg signal extraction by exploiting facial surface orientation. In: Proceedings of the IEEE/CVF Conference on Computer Vision and Pattern Recognition, pp. 2165–2171 (2022)
13. Khan, S.S., Mostafa, N.: Representation biases in remote photoplethysmography datasets-a narrative review. Appl. Comput. Inf. 1–16 (2025)
14. Bondarenko, M., Menon, C., Elgendi, M.: Demographic bias in public remote photoplethysmography datasets. NPJ Dig. Med. **8**(1), 593 (2025)
15. Soleymani, M., Lichtenauer, J., Pun, T., Pantic, M.: A multimodal database for affect recognition and implicit tagging. IEEE Trans. Affect. Comput. **3**(1), 42–55 (2011)
16. Heusch, G., Anjos, A., Marcel, S.: A reproducible study on remote heart rate measurement. arXiv preprint arXiv:1709.00962 (2017)
17. Zhang, Z., et al.: Multimodal spontaneous emotion corpus for human behavior analysis. In: Proceedings of the IEEE Conference on Computer Vision and Pattern Recognition, pp. 3438–3446 (2016)
18. Niu, X., Han, H., Shan, S., Chen, X.: VIPL-HR: a multi-modal database for pulse estimation from less-constrained face video. In: Jawahar, C.V., Li, H., Mori, G., Schindler, K. (eds.) ACCV 2018. LNCS, vol. 11365, pp. 562–576. Springer, Cham (2019). https://doi.org/10.1007/978-3-030-20873-8_36
19. Gudi, A., Bittner, M., Van Gemert, J.: Real-time webcam heart-rate and variability estimation with clean ground truth for evaluation. Appl. Sci. **10**(23), 8630 (2020)
20. Joshi, J., Cho, Y.: Ibvp dataset: Rgb-thermal rppg dataset with high resolution signal quality labels. Electronics **13**(7), 1334 (2024)
21. Revanur, A., Li, Z., Ciftci, U.A., Yin, L., Jeni, L.A.: The first vision for vitals (v4v) challenge for non-contact video-based physiological estimation. In: Proceedings of the IEEE/CVF International Conference on Computer Vision, pp. 2760–2767 (2021)
22. Chari, P., et al.: Diverse r-ppg: camera-based heart rate estimation for diverse subject skin-tones and scenes. arXiv preprint arXiv:2010.12769 (2020)
23. Acharya, B., Saakyan, W., Hammer, B., Drimalla, H.: The reliability of remote photoplethysmography under low illumination and elevated heart rates. npj Dig. Med. (2025)
24. Ma, B., Kaetheeswaran, J., Abedi, A., Lankarany, M., Khan, S.S.: Heart-video – a novel, diverse and inclusive facial video dataset for remote heart rate measurement (2026)
25. Dasari, A., Prakash, S.K.A., Jeni, L.A., Tucker, C.S.: Evaluation of biases in remote photoplethysmography methods. NPJ Dig. Med. **4**(1), 91 (2021)
26. Ba, Y., Wang, Z., Karinca, K.D., Bozkurt, O.D., Kadambi, A.: Style transfer with bio-realistic appearance manipulation for skin-tone inclusive rppg. In: 2022 IEEE International Conference on Computational Photography (ICCP), pp. 1–12. IEEE (2022)
27. Comas, J., Alomar, A., Ruiz, A., Sukno, F.: Physflow: skin tone transfer for remote heart rate estimation through conditional normalizing flows. arXiv preprint arXiv:2407.21519 (2024)

28. Rivest-Hénault, D., et al.: Influential factors in rppg: insights from a diverse and inclusive empirical study. In: 2024 IEEE 20th International Conference on Body Sensor Networks (BSN), pp. 1–4. IEEE (2024)
29. Liu, X., Hill, B., Jiang, Z., Patel, S., McDuff, D.: Efficientphys: enabling simple, fast and accurate camera-based cardiac measurement. In: Proceedings of the IEEE/CVF Winter Conference on Applications of Computer Vision, pp. 5008–5017 (2023)
30. Joshi, J., et al.: Factorizephys: matrix factorization for multidimensional attention in remote physiological sensing. In: Zhang, C. (ed.) Advances in Neural Information Processing Systems, vol. 37, pp. 96607–96639 (2024)
31. Joshi, J., Cho, Y.: Efficient and robust multidimensional attention in remote physiological sensing through target signal constrained factorization. arXiv preprint arXiv:2505.07013 (2025)
32. United Nations High Commissioner for Refugees: Older persons. https://emergency.unhcr.org/protection/persons-risk/older-persons. Accessed 13 June 2025
33. Lou, H., et al.: A large-scale universal evaluation benchmark for face forgery detection. In: Proceedings of the 33rd ACM International Conference on Multimedia, pp. 12592–12599 (2025)
34. Rejón Piña, R.A., Ma, C.: Classification algorithm for skin color (casco): a new tool to measure skin color in social science research. Soc. Sci. Q. **104**(2), 168–179 (2023)
35. Lugaresi, C., et al.: Mediapipe: a framework for perceiving and processing reality. In: Third Workshop on Computer Vision for AR/VR at IEEE Computer Vision and Pattern Recognition (CVPR), vol. 2019 (2019)
36. Liu, X., et al.: rppg-toolbox: deep remote ppg toolbox. Adv. Neural. Inf. Process. Syst. **36**, 68485–68510 (2023)

ACT-A: Acute Concussion Triage Agent for Structured, Guideline-Grounded Concussion Assessment and Referral Support

Tsolmon Gunaabazar[1], Ali Abedi[2(✉)], Tharshini Chandra[1], Mark T. Bayley[1,3], and Shehroz S. Khan[4]

[1] KITE Research Institute, Toronto Rehabilitation Institute, University Health Network, 550 University Avenue, Toronto, ON M5G 2A2, Canada
{tsolmon.gunaabazar,ali.abedi,tharshini.chandra,mark.bayley}@uhn.ca
[2] Peter Munk Cardiac Centre, University Health Network, 585 University Avenue, Toronto, ON M5G 2N2, Canada
[3] Division of Physiatry, Department of Medicine, University of Toronto, 6 Queen's Park Crescent West, Toronto, ON M5S 3H2, Canada
[4] College of Engineering and Technology, American University of the Middle East, 54200 Egaila, Kuwait
shehroz.khan@aum.edu.kw

Abstract. Concussion, or mild traumatic brain injury (mTBI), is a common neurological injury that creates a substantial burden on first-contact healthcare systems. Early triage is clinically important, yet shortage of trained staff, limited access to specialized concussion services, and long wait times can delay assessment and referral. Recent advances in conversational artificial intelligence and large language models have created new opportunities for scalable clinical decision support; however, general-purpose systems remain limited by unstructured inputs, inconsistent outputs, and lack of grounding in clinical guidelines. This paper presents the Acute Concussion Triage Agent (ACT-A), an avatar-based LLM agent designed to support structured and guideline-grounded concussion assessment. ACT-A conducts a structured conversational interview, converts patient responses into organized clinical input, and uses retrieval-augmented generation to ground model reasoning in curated concussion guidelines. The system generates structured outputs, including a concise clinician-facing summary and a next-step referral recommendation. To improve consistency with task-specific clinical decision boundaries, in-context learning was incorporated using labeled examples of patient profiles and corresponding referral dispositions. The system was evaluated using synthetic concussion profiles designed to represent clinically realistic mTBI presentations across different mechanisms of injury, symptom patterns, risk factors, and acuity levels. Results showed that retrieval grounding improved identification of higher-risk cases and produced more clinically specific summaries, while in-context

T. Gunaabazar and A. Abedi—These authors contributed equally to this work.

S. S. Khan et al. (Eds.): IJCAI 2026, CCIS 3127, pp. 152–163, 2027.
https://doi.org/10.1007/978-981-92-5190-2_11

learning further improved alignment between model recommendations and expected referral decisions. These findings suggest that combining avatar-based interaction, structured clinical inputs, retrieval-augmented reasoning, and in-context learning can support safer and more consistent AI-assisted mTBI triage. ACT-A provides a foundation for evidence-based clinical decision support in first-contact concussion care, pending future validation with real-world patient data.

Keywords: concussion · mild traumatic brain injury · conversational avatar · LLM agents · retrieval-augmented generation · in-context learning · clinical decision support

1 Introduction

Concussion, also referred to as mild traumatic brain injury (mTBI), is a common neurological condition caused by biomechanical forces such as falls, motor vehicle collisions, workplace injuries, and sports-related impacts [1]. Although many patients recover within weeks, a substantial subset experience persistent post-concussion symptoms, including headache, dizziness, fatigue, impaired concentration, emotional symptoms, and reduced functional capacity [2,3]. Early assessment and appropriate referral are therefore central to concussion management, as delayed evaluation has been associated with persistent symptoms during recovery [4]. However, first-contact care settings often face shortages of trained staff, limited access to specialized concussion services, and long wait times for assessment. These barriers can delay triage, increase variability in clinical decision-making, and limit timely access to appropriate follow-up care.

Current concussion assessment relies heavily on clinician-led history-taking and patient self-report, often supported by standardized symptom inventories and clinical practice guidelines for mTBI and persistent symptoms [5,6]. The guideline work by Marshall, Bayley, and colleagues emphasizes structured assessment, symptom monitoring, and evidence-informed management for adults with concussion/mTBI and persistent symptoms [6]. While these approaches are clinically useful, they can be time-intensive, subjective, and difficult to apply consistently in busy emergency, primary care, or community settings. Patient-reported symptoms may also be incomplete or difficult to interpret during the acute phase of injury, particularly when cognitive, emotional, vestibular, or oculomotor symptoms are present.

Recent advances in conversational artificial intelligence and large language models (LLMs) have created new opportunities for AI-assisted clinical triage and decision support [7,8]. LLM agents can support structured interactions, synthesize patient information, and generate summaries that may reduce documentation burden in clinical workflows. In parallel, conversational avatar interfaces can provide a more guided and natural interaction than static questionnaires or text-only chatbots, which may be particularly relevant for patient-facing symptom collection [9,10]. However, general-purpose LLM systems remain limited for safety-critical clinical triage because their outputs may be inconsistent, difficult to audit, and insufficiently grounded in clinical evidence or domain-specific guidelines [11,12].

Retrieval-augmented generation (RAG) provides a technical approach for addressing this limitation by combining the LLM's parametric knowledge with retrieved external documents during inference [12]. Evidence- and guideline-grounded reasoning refers to the use of retrieved concussion guideline content to support the model's clinical summary and referral recommendation, rather than relying solely on pretrained model knowledge. This design is intended to improve alignment with documented clinical criteria, increase transparency, and reduce the risk of unsupported or inconsistent recommendations. In-context learning (ICL) provides a complementary mechanism by supplying labeled examples within the prompt, allowing the model to better align its outputs with task-specific referral categories without additional model training [13].

This paper presents the Acute Concussion Triage Agent (ACT-A), a conversational avatar-based LLM agent designed to support structured and guideline-grounded concussion assessment. ACT-A conducts a structured concussion interview through an avatar-based interface, converts patient responses into organized clinical input, and generates structured outputs for clinician review. These outputs include a concise clinician-facing summary and a next-step referral recommendation. ACT-A is designed as a clinical decision-support system rather than an autonomous diagnostic or referral tool.

ACT-A is evaluated using synthetic mTBI profiles generated as structured variants of 10 real-world concussion patient presentations. These variants were designed to preserve clinically realistic patterns while systematically varying triage-relevant features, including mechanism of injury, symptom clusters, symptom severity, time since injury, red flag indicators, prior concussion history, and relevant medical context.

The main contributions are the development of an avatar-based LLM agent for structured concussion assessment, the integration of RAG for guideline-grounded clinical reasoning, the use of ICL to improve alignment with expected referral decisions, and an evaluation framework based on clinically realistic synthetic variants derived from real-world concussion presentations. The system is positioned as an objective second opinion to support clinician decision-making; final decisions regarding referral or next-step care remain the responsibility of qualified clinicians.

2 Literature Review

Digital tools for patient triage have evolved from static symptom checkers toward conversational systems that can collect patient information, reason over clinical presentations, and recommend levels of care. Early digital symptom checkers demonstrated the feasibility of large-scale, patient-facing symptom assessment and care navigation, but they often relied on structured questionnaires and rule-based logic rather than adaptive clinical dialogue [14,15]. More recent chatbot-based systems have been evaluated in clinical domains such as ophthalmology and emergency medicine, showing that generative AI systems can produce triage decisions that are sometimes comparable to clinicians or trainees in controlled vignette-based settings [16,17]. These studies suggest that conversational AI may help expand access to preliminary assessment, but they also highlight the need for careful evaluation because triage requires sensitivity to incomplete information, high-risk symptoms, and escalation thresholds.

LLMs have accelerated this shift by enabling more flexible patient-facing dialogue and clinical reasoning over free-text information. In emergency medicine, comparative evaluations have shown that LLMs can assign triage acuity and reason over patient presentations, although performance varies by model, task formulation, and clinical context [17,18]. Recent work using real emergency department cases further suggests that advanced reasoning models may perform well on text-based emergency triage and clinical decision tasks, supporting their potential role as second-opinion systems rather than replacements for clinicians [19]. However, these systems remain limited when patient information is incomplete or when recommendations are generated without transparent grounding in clinical guidance. This limitation is particularly important in first-contact triage, where under-triage may delay urgent care and overconfident recommendations may create safety risks.

Agent-based approaches have been proposed to improve the reliability and structure of LLM-based triage. TriageAgent, for example, uses a heterogeneous multi-agent framework for clinical triage, combining role-specific agents, multi-round discussion, confidence estimation, early stopping, and retrieval from emergency severity guidance to improve triage classification [20]. Similarly, Rashidian and colleagues developed conversational patient triage agents evaluated through simulated encounters derived from real-world electronic health record data, demonstrating a workflow in which patient simulators and triage agents can be used to develop and test scalable AI triage systems before clinical deployment [8]. These studies are directly relevant to ACT-A because they frame triage as an interactive reasoning process rather than a single-step classification task. They also show that simulated or synthetic patient encounters can provide a practical intermediate evaluation strategy when prospective clinical testing is not yet available.

Retrieval-grounded approaches have also become increasingly important for clinical triage. RAG-based systems retrieve external clinical evidence during inference, allowing the model to condition its output on explicit guidelines or reference materials rather than relying only on internal model knowledge [12]. In emergency triage, multi-evidence RAG has been evaluated as a way to improve clinical reasoning by integrating multiple retrieved evidence sources into the model's decision process [11]. This direction is important for patient-facing triage systems because clinical recommendations must be auditable, evidence-aligned, and conservative enough to avoid missing high-acuity presentations.

Avatar-based systems represent a related but less extensively evaluated direction for triage. Existing virtual human and avatar applications have mainly been studied for patient engagement, mental health assessment, education, and communication training rather than formal triage decision-making [10,21]. Nevertheless, avatars may provide a useful interface for triage by making symptom collection more guided, accessible, and conversational than text-only systems. The remaining gap is therefore not simply the use of an avatar, chatbot, or LLM in isolation, but the integration of a browser-accessible conversational avatar with structured patient input, guideline-grounded LLM reasoning, and standardized referral outputs. ACT-A addresses this gap by combining an avatar-based assessment workflow with RAG and in-context learning to support structured concussion triage as an objective second opinion for clinicians.

3 Methods

This section describes the ACT-A system architecture, the conversational avatar-based assessment workflow, the guideline-grounded RAG pipeline, the ICL strategy, the generation of synthetic patient profiles, and the clinical decision-support framing used to position the system as an assistive tool for clinician review.

3.1 ACT-A System Overview

ACT-A was developed as a conversational avatar-based LLM agent for structured and guideline-grounded concussion assessment. The system was designed to support first-contact triage for patients with suspected concussion or mTBI. The ACT-A workflow consists of four main steps: structured patient interaction through a conversational avatar, conversion of patient responses into organized clinical input, guideline-grounded LLM reasoning, and generation of structured outputs for clinician review.

ACT-A was designed as a clinical decision-support system rather than an autonomous diagnostic or referral tool. The system output consists of a concise clinician-facing summary and a next-step care recommendation. The recommendation is selected from predefined referral categories, including (1) discharge home, (2) follow-up with primary care, (3) referral to a concussion clinic, and (4) referral to the emergency department (ED). This structured output format was intended to reduce variability in documentation, support consistent triage reasoning, and provide clinicians with an objective second opinion.

3.2 Conversational Avatar-Based Assessment Workflow

The patient-facing component of ACT-A was implemented as a browser-accessible web application containing a 3D animated clinician avatar, enabling use across internet-connected computers, laptops, tablets, and mobile devices. An illustration of ACT-A running within a web browser is presented in Fig. 1. The digital clinician avatar conducts a structured concussion history interview using predefined clinical questions. During the assessment, patient speech is converted into text through the browser-based interaction layer on the local device, and the resulting textual data are transferred to the Microsoft Azure cloud environment for downstream processing.

The transferred data are converted into structured clinical input for the LLM agent. This structured input includes clinically relevant information such as mechanism of injury, time since injury, symptom profile, symptom severity, prior concussion history, medication use, and red flag indicators. The structured textual profile is then processed by the RAG, LLM, and ICL framework to generate the clinician-facing summary and referral recommendation. This design allows the LLM agent to reason over organized clinical fields rather than unstructured free-text dialogue alone. The purpose of this workflow is to support systematic symptom collection before clinician review, particularly in settings where staff shortages and long wait times may limit the consistency or timeliness of early concussion triage.

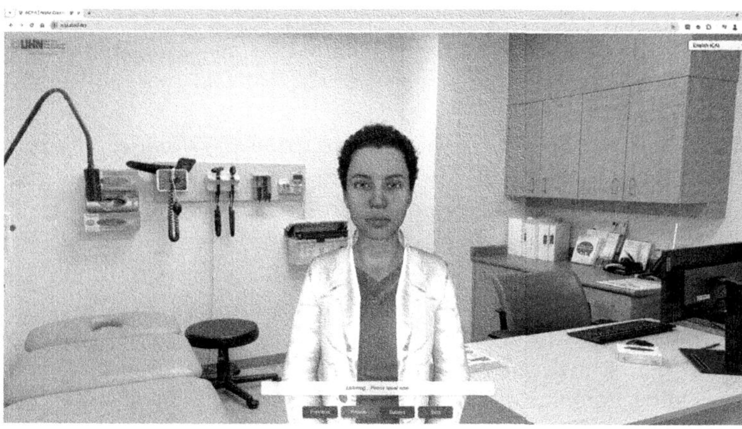

Fig. 1. Browser-based ACT-A interface showing the 3D animated clinician avatar used for structured concussion assessment.

3.3 Guideline-Grounded RAG Pipeline

RAG was used to ground ACT-A outputs in external clinical evidence. In this study, evidence- and guideline-grounded reasoning refers to the use of retrieved concussion-related clinical guidance to support the model's clinical summary and referral recommendation, rather than relying only on pretrained model knowledge [12,22]. This approach is particularly relevant for clinical triage, where unsupported or inconsistent recommendations may create patient-safety risks. The evidence corpus was composed of curated concussion guidance and symptom-assessment materials, including clinical practice guidelines for concussion/mTBI and persistent symptoms, as well as standardized symptom-assessment literature [5,6]. These documents were used as the grounding source for the LLM agent.

The RAG pipeline was implemented by segmenting the guideline corpus into overlapping text chunks, embedding each chunk into a vector representation, and storing the resulting vectors in a similarity-search index. Vector retrieval was performed using FAISS [23]. Sentence-transformer embeddings were used to represent text chunks and patient profiles in the same semantic space [24]. At inference time, the structured patient profile was converted into a retrieval query, and the most relevant guideline passages were retrieved and inserted into the LLM prompt. The LLM was instructed to use the retrieved evidence when generating two structured outputs: a concise clinician-facing summary and exactly one referral recommendation.

3.4 In-Context Learning Strategy

ICL was used to improve alignment between ACT-A outputs and task-specific referral categories. In this study, labeled examples of patient profiles and corresponding referral dispositions were included in the prompt to demonstrate the expected mapping between structured clinical features and next-step care recommendations [13].

The in-context examples were designed to expose the model to representative decision boundaries across the four referral categories. This was particularly important because concussion triage often depends on combinations of factors rather than a single symptom. For example, symptom severity, time since injury, red flag indicators, prior concussion history, and mechanism of injury may jointly influence whether primary care follow-up, concussion clinic referral, or emergency assessment is most appropriate. By providing labeled examples, the prompt was structured to help the LLM agent apply consistent reasoning to new profiles while maintaining guideline-grounded evidence through the RAG context.

The final LLM prompt therefore contained three components: the structured patient profile, the retrieved guideline evidence, and selected labeled examples for ICL. The model was then instructed to produce two structured outputs: a concise clinician-facing summary and exactly one referral recommendation. This combination of RAG and ICL was intended to support both evidence grounding and task-specific decision alignment.

3.5 Synthetic Patient Profile Generation

ACT-A was evaluated using synthetic mTBI profiles generated as structured variants of 10 real-world concussion patient presentations. The term "synthetic variants" refers to systematically modified versions of real-world presentation patterns, rather than fully artificial or randomly generated cases. This approach was used to preserve clinically realistic symptom patterns while allowing controlled variation across triage-relevant features. Synthetic profiles were used because real patient data were not yet available for end-to-end evaluation, and because controlled profiles enabled systematic testing of the LLM agent across a broader range of clinical scenarios before real-world deployment.

The synthetic profiles varied across demographic and clinical dimensions, including age, sex, mechanism of injury, time since injury, symptom clusters, symptom severity, red flag indicators, prior concussion history, medication use, and relevant medical context. Symptom profiles were designed to reflect common concussion presentations, including headache, dizziness, fatigue, cognitive difficulty, light or noise sensitivity, emotional symptoms, and vestibular or oculomotor complaints. Referral labels were assigned using a rule-based framework informed by concussion triage logic and clinical guidance for mTBI and persistent symptoms [2,6].

Each synthetic profile was used as a structured input to the ACT-A LLM agent. The model generated a clinician-facing summary and one referral recommendation for each profile. The assigned referral label served as the reference disposition for evaluating whether the model output was aligned with the expected triage decision.

4 Results

This section first summarizes the baseline characteristics of the synthetic patient profiles used for evaluation, followed by a description of the evaluation criteria covering both metric-based assessment of referral recommendations and qualitative assessment of generated summaries. RAG and No-RAG performance was evaluated on the full set of 256 synthetic patient profiles, where No-RAG refers to standalone LLM inference based only on the model's internal knowledge without access to retrieved guideline passages or other external evidence. ICL performance was evaluated on a held-out subset of 200 profiles, with the remaining 56 profiles reserved as labeled examples for in-context prompting. Finally, generated clinical summaries from the RAG and No-RAG conditions were compared using a blinded LLM-as-judge analysis across all 256 profiles.

4.1 Baseline Characteristics of Synthetic Patient Profiles

The full synthetic profile set included 256 profiles, of which 200 were used as held-out evaluation profiles and 56 were reserved as labeled examples for ICL. The profiles covered an age range of 13–85 years, with a mean age of 36.5 years, median age of 33.5 years, and standard deviation of 17.92 years. Gender identity was distributed as women 50.8%, men 41.4%, non-binary 3.1%, trans women 2.3%, trans men 0.8%, and prefer not to say 1.6%. Sex was distributed as female 52.7%, male 44.9%, and intersex 2.3%.

4.2 Evaluation Metrics

Quantitative performance was assessed using accuracy, macro F1 score, and ED recall. Accuracy measured the proportion of profiles for which the model recommendation matched the reference referral label. Macro F1 was used to summarize performance across the four referral classes while accounting for class-level imbalance. ED recall measured the proportion of true emergency referral cases correctly identified by the model and was emphasized as a safety-relevant measure because missed high-acuity cases carry greater clinical risk than conservative over-referral. In addition to metric-based evaluation, generated clinical summaries were assessed using a blinded LLM-as-judge analysis to compare the clinical usefulness, specificity, and appropriateness of RAG and No-RAG outputs.

4.3 RAG Versus No-RAG Performance

Table 1 presents the performance of ACT-A under No-RAG and RAG conditions across the 256 synthetic patient profiles. Across all models, RAG improved ED recall, indicating better identification of high-acuity cases. The largest improvement was observed for Claude Sonnet 4.6, where ED recall increased from 44.3% to 65.6%, accompanied by modest improvements in accuracy and macro F1. GPT-4o also showed improved ED recall with RAG, although accuracy decreased from 45.3% to 42.9% and macro F1 changed minimally from 0.416 to 0.414. GPT-5.4 maintained the same overall accuracy under both conditions, showed a small improvement in ED recall, and had a modest decrease in macro F1 from 0.346 to 0.333. Overall, these findings suggest that guideline-grounded retrieval shifted model behavior toward safer, higher-acuity recommendations, although this benefit was not consistently accompanied by improved overall accuracy or macro F1 score.

Table 1. RAG versus No-RAG performance on 256 patient profiles.

Model	Setup	Accuracy	Macro F1	ED Recall
GPT-4o	No-RAG	45.3%	0.416	65.6%
GPT-4o	RAG	42.9%	0.414	70.5%
Claude Sonnet 4.6	No-RAG	37.5%	0.323	44.3%
Claude Sonnet 4.6	RAG	38.7%	0.333	65.6%
GPT-5.4	No-RAG	41.4%	0.346	82.0%
GPT-5.4	RAG	41.4%	0.333	83.6%

4.4 Effect of In-Context Learning

Table 2 summarizes model performance under the ICL condition using 200 held-out synthetic patient profiles. Compared with the RAG and No-RAG results reported above, ICL improved overall referral classification performance on the 200-profile held-out subset, particularly for accuracy and macro F1 score. Claude Sonnet 4.6 achieved the highest overall accuracy and macro F1 score, indicating the strongest balance across referral categories. GPT-5.4 also showed improved overall performance compared with its RAG and No-RAG results, although its

ED recall was lower than in the RAG-only condition. Grok 4.2 achieved the highest ED recall among the ICL models, suggesting stronger sensitivity for high-acuity cases. Overall, these findings suggest that ICL improved alignment with expected referral decisions, while model-specific tradeoffs remained between overall classification accuracy and detection of ED referral cases.

Table 2. In-context learning performance on 200 held-out synthetic patient profiles.

Model	Accuracy	Macro F1	ED Recall
Claude Sonnet 4.6	55.5%	0.554	64.7%
Grok 4.2	51.0%	0.485	78.4%
Claude Opus 4.6	50.5%	0.501	47.1%
GPT-5.4	48.5%	0.478	62.7%

4.5 LLM-As-Judge Evaluation of Generated Clinical Summaries

Generated clinical summaries were evaluated using a blinded LLM-as-judge analysis. For each of the 256 synthetic patient profiles, the judge was presented with two candidate summaries: one generated under the RAG condition and one generated under the No-RAG condition. The source of each summary was hidden and randomized. Summaries were compared based on consistency with the patient profile, clinical specificity, appropriateness of the recommended disposition, and clarity for clinical decision-making.

In the blinded evaluation, RAG-generated summaries were preferred in 75 of 256 cases (29.3%), compared with 49 cases (19.1%) for No-RAG summaries. In 132 cases (51.6%), both summaries were judged equivalent. These findings indicate that RAG provided a qualitative benefit in a subset of cases, particularly when retrieved guideline content supported more clinically specific reasoning. However, the high proportion of ties suggests that both RAG and No-RAG conditions often generated clinically usable summaries.

5 Conclusion and Future Work

This paper presented ACT-A as a conversational avatar-based LLM system for structured and guideline-grounded concussion assessment. ACT-A generates clinician-facing summaries and referral recommendations using RAG and ICL. Evaluation on synthetic mTBI profiles showed that RAG improved ED recall, suggesting better sensitivity for high-acuity cases, while ICL improved alignment with expected referral decisions. LLM-as-judge evaluation also showed that RAG summaries were preferred more often than No-RAG summaries. Future work should evaluate ACT-A using data collected from real-world concussion patients and assess its performance in clinical workflows. Further work should also examine usability for clinicians and patients, improve model calibration, update the RAG knowledge base, and develop a multi-agent extension in which specialized agents coordinate symptom collection, evidence retrieval, risk assessment, and recommendation verification.

References

1. Katz, D.I., Cohen, S.I., Alexander, M.P.: Mild traumatic brain injury. Handb. Clin. Neurol. **127**, 131–156 (2015)
2. Dwyer, B., Katz, D.I.: Postconcussion syndrome. Handb. Clinical Neurol. **158**, 163–178 (2018)
3. Sheldrake, E., et al.: Mental health outcomes across the lifespan in individuals with persistent post-concussion symptoms: a scoping review. Front. Neurol. **13**, 850590 (2022)
4. Bunt, S.C., et al.: Early vs. delayed evaluation and persisting concussion symptoms during recovery in adults. Clin. Neuropsychol. **37**(7), 1410–1427 (2023)
5. Karr, J.E., Zuccato, B.G., Ingram, E.O., McAuley, T.L., Merker, B., Abeare, C.A.: The post-concussion symptom scale: normative data for adolescent student-athletes stratified by gender and preexisting conditions. Am. J. Sports Med. **51**(1), 225–236 (2023)
6. Marshall, S., et al.: Updated clinical practice guidelines for concussion/mild traumatic brain injury and persistent symptoms. Brain Inj. **29**(6), 688–700 (2015)
7. Anisha, S.A., Sen, A., Bain, C.: Evaluating the potential and pitfalls of ai-powered conversational agents as humanlike virtual health carers in the remote management of noncommunicable diseases: scoping review. J. Med. Internet Res. **26**, 56114 (2024)
8. Rashidian, S., et al.: Ai agents for conversational patient triage: preliminary simulation-based evaluation with real-world ehr data. arXiv preprint arXiv:2506.04032 (2025)
9. Heyselaar, E., Hagoort, P., Segaert, K.: In dialogue with an avatar, language behavior is identical to dialogue with a human partner. Behav. Res. Methods **49**(1), 46–60 (2017)
10. DeTore, N.R., Balogun-Mwangi, O., Eberlin, E.S., Dokholyan, K.N., Rizzo, A., Holt, D.J.: An artificial intelligence-based virtual human avatar application to assess the mental health of health care professionals: a validation study. J. Med. Extended Reality **1**(1), 2024 (2024)
11. Wong, H.S., Wong, T.K.: Multi-evidence clinical reasoning with retrieval-augmented generation for emergency triage: retrospective evaluation study. JMIR Med. Inf. **14**(1), 82026 (2026)
12. Lewis, P., et al.: Retrieval-augmented generation for knowledge-intensive nlp tasks. Adv. Neural. Inf. Process. Syst. **33**, 9459–9474 (2020)
13. Brown, T., et al.: Language models are few-shot learners. Adv. Neural. Inf. Process. Syst. **33**, 1877–1901 (2020)
14. Morse, K.E., Ostberg, N.P., Jones, V.G., Chan, A.S.: Use characteristics and triage acuity of a digital symptom checker in a large integrated health system: population-based descriptive study. J. Med. Internet Res. **22**(11), 20549 (2020)
15. Clark, M., Bailey, S.: Chatbots in health care: connecting patients to information. Canad. J. Health Technol. **4**(1) (2024)

16. Lyons, R.J.: Artificial intelligence chatbot performance in triage of ophthalmic conditions. Can. J. Ophthalmol. (2024)
17. Masanneck, L., et al.: Triage performance across large language models, chatgpt, and untrained doctors in emergency medicine: Comparative study. J. Med. Internet Res. **26**, 53297 (2024)
18. Nedos, I., et al.: Is artificial intelligence ready for emergency department triage? J. Clin. Med. (2026)
19. Brodeur, P.G.: Performance of a large language model on the reasoning tasks of emergency medicine physicians. Science (2026)
20. Lu, M., Ho, B., Ren, D., Wang, X.: Triageagent: towards better multi-agents collaborations for large language model-based clinical triage. In: Findings of the Association for Computational Linguistics: EMNLP 2024, pp. 5747–5764 (2024)
21. Gutiérrez Maquilón, R., et al.: Integrating gpt-based ai into virtual patients to facilitate communication training: development and evaluation study. JMIR Form. Res. **8**, 58623 (2024)
22. Abedi, A., Chu, C.H., Khan, S.S.: Evidence-informed guidance on cannabidiol use in older adults: development and evaluation of retrieval-augmented large language models. preprint (2026)
23. Johnson, J., Douze, M., Jégou, H.: Billion-scale similarity search with gpus. IEEE Trans. Big Data **7**(3), 535–547 (2019)
24. Reimers, N., Gurevych, I.: Sentence-bert: sentence embeddings using siamese bert-networks. In: Proceedings of the 2019 Conference on Empirical Methods in Natural Language Processing and the 9th International Joint Conference on Natural Language Processing (EMNLP-IJCNLP), pp. 3982–3992 (2019)

A Markerless AI Goniometer for Ankle Dorsiflexion Measurement: Algorithm Design, Model Evaluation, and Rehabilitation Implications for Elders

Naomi Unkelos-Shpigel[1](✉), Ron Kroitoro[1], Sheli Zisman[1], and Einat Ravid[2]

[1] Department of Software Engineering, Braude College of Engineering, Karmiel, Israel
naomius@braude.ac.il

[2] Department of Electrical Engineering, Braude College of Engineering, Karmiel, Israel

Abstract. Accurate measurement of ankle dorsiflexion range of motion (ROM) is fundamental to rehabilitation and orthopedic practice, yet the gold-standard universal goniometer is widely abandoned in daily clinical use due to poor inter-rater reliability. This paper presents an AI-based markerless image-analysis system for automated dorsiflexion measurement, developed through a clinician needs assessment and evaluated as a trained deep-learning model. The system employs DeepLabCut with a ResNet-50 backbone to detect three anatomical keypoints (tibial midline, lateral malleolus, fifth metatarsal head) from standard smartphone photographs and computes the dorsiflexion angle via vector-based geometric analysis. The model (DLC_Resnet50_google-pixOct5shuffle1) was trained on a dedicated clinical dataset of 369 images from 16 participants (80/20 train/test split) and achieved a training RMSE of 6.07 px (5.27 px at p-cutoff), training mAP of 97.54%, test mAP of 94.48%, and test RMSE of 7.52 px at p-cutoff. Geometric analysis of the imaging setup confirms that this localization error corresponds to a maximum angle error of approximately 1.7°, well within the $\leq 5°$ clinical requirement. A clinical validation study comparing the system against the universal goniometer and a protractor-mounted reference device is underway. The system is designed for clinician-in-the-loop deployment in rehabilitation settings, with longitudinal tracking and EHR integration capabilities relevant to aging and post-injury monitoring.

Keywords: ankle dorsiflexion · range of motion · markerless pose estimation · DeepLabCut · ResNet-50 · key point detection · rehabilitation

1 Introduction

Ankle dorsiflexion, the movement of bringing the toes toward the lower leg, is a fundamental parameter in rehabilitation and orthopedic assessment. It enables forward propulsion during gait and ground clearance during the swing phase [3]. Restricted dorsiflexion ROM is a frequent consequence of injury and aging, and is associated with altered joint kinematics, reduced stability, increased fall risk, and impaired quality of life [1, 4, 6, 9]. In

S. S. Khan et al. (Eds.): IJCAI 2026, CCIS 3127, pp. 164–173, 2027.
https://doi.org/10.1007/978-981-92-5190-2_12

older adults, these effects compound with age-related reductions in muscle strength and proprioception, making reliable and longitudinal ROM monitoring particularly critical [12].

The gold-standard universal goniometer (UG) is known to produce poor inter-rater reliability: clinicians interviewed in our needs assessment described routinely abandoning formal measurement in favor of visual estimation, citing inconsistency when the device is moved even slightly. Yet the downstream clinical costs of measurement error are concrete: inappropriate surgical decisions, suboptimal rehabilitation trajectories, and missed mobility restrictions. This gap motivates a fundamentally different approach.

Recent advances in markerless pose estimation offer a viable path [11]. DeepLabCut [11], a framework originally developed for animal behavior analysis, enables training of customized key point detection models on task-specific datasets with high spatial accuracy. Applied to clinical ankle images, such a model can detect anatomical landmarks automatically, compute the dorsiflexion angle geometrically, and report a result in seconds, without requiring markers, sensors, or specialist equipment beyond a smartphone. This paper contributes: (1) a clinician needs assessment documenting the failure modes of current measurement practice and the requirements for a digital replacement; (2) the design and implementation of a six-stage AI pipeline for ankle dorsiflexion measurement; (3) model training and evaluation results on a dedicated clinical dataset; and (4) a planned clinical validation protocol. The work addresses the themes of AI for rehabilitation monitoring, clinician-in-the-loop systems, and trustworthy AI deployment in care environments.

2 Related Work

2.1 Limitations of Goniometric Measurement

Rome and Cowieson [14] demonstrated substantial inter-rater variability for universal goniometer measurements of ankle dorsiflexion. Youdas et al. [18] confirmed that both goniometric and visual estimates are unreliable, with errors sufficient to misattribute noise to clinical change. Manual handling amplifies error further, and the inability to log measurements across sessions prevents longitudinal monitoring. Collins et al. [6] systematically reviewed reliability across clinical techniques and concluded that no manual method consistently achieves the accuracy required for clinical decision-making.

2.2 Markerless Pose Estimation for Clinical Measurement

Deep learning frameworks for pose estimation have achieved high accuracy in biomechanical research [10, 11]. DeepLabCut [11] enables training of customized models on small, annotated clinical datasets, detecting user-defined anatomical landmarks with subpixel accuracy using ResNet-50 as a feature extractor. Kidzinski et al. [9] demonstrated that deep neural networks enable quantitative movement analysis from single-camera video. Roggio et al. [13] surveyed pose estimation models in human movement analysis and concluded that task-specific models outperform general-purpose frameworks for fine-grained landmark localization. In our comparison, MediaPipe Pose and YOLO Key points both produced insufficient precision for ankle landmarks, motivating the custom DeepLabCut approach.

2.3 AI for Rehabilitation and Aging

AI-assisted ROM measurement is part of a broader movement toward continuous, non-invasive monitoring in rehabilitation and aging care [5, 15]. Automated measurement tools that integrate with clinical workflows reduce examiner burden, enable longitudinal data collection, and support data-driven rehabilitation programs. Asan et al. [2] showed that clinician trust in AI tools depends on transparency and verifiability, findings that directly informed our interface design. Learning from limited annotated data, a key challenge in clinical AI, is addressed here through task-specific training on a dedicated ankle dataset [15].

3 Clinician Needs Assessment

3.1 Expert Interviews

We conducted semi-structured interviews with two orthopedic specialists at a major hospital: a senior orthopedic surgeon (head of department) and an orthopedic resident, both with several years of daily goniometer experience. Each interview lasted approximately 30 min. Key findings:

"To be honest, we're so busy that we often just estimate the angle 'by eye.' Even when we do use the goniometer, if you move it just a tiny bit, the result changes completely. It's just not consistent." (Senior Orthopedic Surgeon, Head of Department, Hospital X).

"During dorsiflexion measurement, it is very difficult to accurately identify the correct measurement point, especially when the foot is swollen. Ankle anatomy varies between patients, and the measurement relies heavily on visual estimation and manual goniometry. In practice, this often results in inconsistent measurements and a form of angle 'guessing.'" (Orthopedic Resident, Hospital X).

3.2 Survey Findings

A structured anonymous survey was administered digitally to nine practitioners (six physiotherapists, two orthopedic surgeons, one podiatric surgeon), all with over five years of clinical experience [15]. Seven open-ended questions addressed the clinical consequences of measurement error, current goniometer usage, perceived accuracy, desired features, and adoption intent. Thematic analysis revealed six recurring categories: clinical consequences of inaccurate measurement, rehabilitation impact, perceived accuracy, current goniometer usage patterns, desired simplicity, and desired data integration [15]. Participants described choosing inappropriate surgical interventions and missing opportunities to recommend stretching as direct consequences of measurement error and reported rarely or never using the goniometer in daily practice, relying instead on visual estimation. The results revealed a clear knowledge-practice gap: participants were acutely aware of the downstream clinical costs of inaccurate ROM assessment, yet most had abandoned formal goniometric measurement in daily practice, citing unreliable results and procedural burden. A viable replacement must therefore be not only more accurate than the universal goniometer, but meaningfully faster and less disruptive to clinical workflows.

4 The Proposed System: Algorithm and Architecture

4.1 System Overview and Model Selection

The system accepts two standard photographs of the ankle, one in neutral position and one at maximum dorsiflexion, and requires no markers, sensors, or equipment beyond a smartphone. Three anatomical key points are detected automatically (tibial midline, lateral malleolus, fifth metatarsal head) and the dorsiflexion angle is computed via vector-based geometric analysis. The ROM delta (neutral angle minus flexion angle) constitutes the final measurement output.

We evaluated three candidate approaches before selecting the final algorithm. Manual key point annotation combined with classical geometry was rejected due to high inter-rater variability and the requirement for continuous human involvement. Pre-trained general-purpose pose estimation models (MediaPipe Pose, YOLO Key points, OpenPose) were evaluated directly on ankle images and rejected because they lack the precision required for fine-grained ankle landmark localization: all three produced unacceptable landmark localization errors on the lateral malleolus and fifth metatarsal head under clinical imaging conditions. A custom DeepLabCut model trained on a dedicated ankle dataset was selected, as it enables precise definition of task-specific key points and achieves high accuracy even with limited annotated training data [8].

4.2 DeepLabCut with ResNet-50: Architecture

DeepLabCut [11] is a transfer-learning-based pose estimation framework that fine-tunes a pretrained convolutional neural network backbone on a small, user-annotated dataset to detect arbitrary user-defined key points. The framework outputs per-key point (x, y) coordinates and a confidence score (likelihood), enabling downstream filtering by a confidence cutoff (p-cutoff) to reject uncertain predictions.

ResNet-50 [He et al., 2016] serves as the feature extraction backbone. Its residual architecture, with skip connections that propagate gradients across 50 layers, is particularly effective for learning spatially precise features from small clinical datasets. The skip connections mitigate the vanishing-gradient problem, preserve fine-grained spatial information from early layers, and stabilize training, all of which are critical for accurate sub-pixel landmark localization in ankle images where target structures span only a few pixels. The pretrained weights (initialized from ImageNet) allow the model to converge reliably with the limited annotated data available in a clinical collection pipeline.

The trained model, designated DLC_Resnet50_google-pixOct5shuffle1, was selected as the best snapshot at training iteration 120, based on validation performance. The model detects all three ankle key points in a single forward pass, returning pixel coordinates and per-key point confidence scores that are fed into the downstream angle computation module.

4.3 Dataset and Training

A dedicated dataset was collected from 16 healthy adult participants (10 male, 6 female; undergraduate students, mean age 25 years, range 20–30), with no active ankle injury or musculoskeletal condition at the time of data collection. All participants provided written informed consent prior to image acquisition. Each participant was photographed in both neutral and maximum dorsiflexion positions, yielding 369 images in total (approximately 23 images per participant). Images were captured using an iPhone 14 (12 MP rear camera, native resolution 4032 × 3024 px) mounted on a fixed tripod at a camera-to-subject distance of 1.2 m, with the participant seated, knee flexed at approximately 90 degrees, lower leg vertically, against a white cardboard background under standardized indoor lighting. All three anatomical key points were manually annotated in each image by a single trained annotator, with anatomical landmark positions defined in consultation with an orthopedic specialist. The dataset was partitioned into a training set (80%, 295 images) and a held-out test set (20%, 74 images), with no participant overlap between splits. Table 1 summarizes the dataset characteristics.

Table 1. Dataset characteristics.

Parameter	Value
Acquisition setup	White background, standardized indoor lighting, knee at ~ 90°
Annotated key points per image	3 (tibial midline, lateral malleolus, fifth metatarsal head)
Annotation	Manual, single trained annotator
Train/test split	80%/20% (295/74 images, no participant overlap)
Annotated key points per image	3 (tibial midline, lateral malleolus, fifth metatarsal head)

Challenges during dataset creation included: (1) image acquisition variability in lighting, camera angle, and background; (2) patient-related variability including differences in body composition, calf-muscle structure, and anatomical variation that can obscure the lateral malleolus and fifth metatarsal; (3) the labor-intensive nature of manual annotation, requiring careful examination of each image; and (4) the need for anatomical calibration markers during initial data collection to guide the annotator. The model was trained using DeepLabCut's standard pipeline with data augmentation, and the best-performing snapshot was selected based on validation RMSE at the chosen p-cutoff.

4.4 Angle Computation Algorithm

Once key points are detected, the dorsiflexion angle is computed using vector-based geometric analysis. Let $K_1 = (x_1, y_1)$ denote the tibial midline key point, $K_2 = (x_2, y_2)$ the lateral malleolus, and $K_3 = (x_3, y_3)$ the fifth metatarsal head. Two vectors are defined: the shank vector $v_1 = K_1 - K_2$ (tibial axis) and the foot vector $v_2 = K_3 - K_2$ (metatarsal axis). The included angle θ is computed as:

$$\theta = arccos(v_1 \cdot v_2/(|v_1| \cdot |v_2|))$$

The dorsiflexion ROM delta is then: $\Delta ROM = \theta\_neutral - \theta\_flexion$, where the neutral angle establishes the individual's anatomical baseline. Key points with confidence below the p-cutoff threshold are flagged and excluded from the calculation; the system returns an uncertainty warning rather than an unreliable angle. This design preserves clinical safety by making uncertainty explicit rather than hiding it in a spuriously precise numerical output.

4.5 Model Evaluation Results

The model was evaluated on training and held-out test sets [11]. Figure 1 illustrates the system output for a representative image. The three detected key points are shown overlaid on the ankle image: the tibial midline (upper marker), lateral malleolus (middle marker), and fifth metatarsal head (distal marker). The two vectors derived from these points define the shank and foot axes, and the included angle constitutes the dorsiflexion measurement. The computed angle and per-key point confidence scores are displayed alongside the visualization:

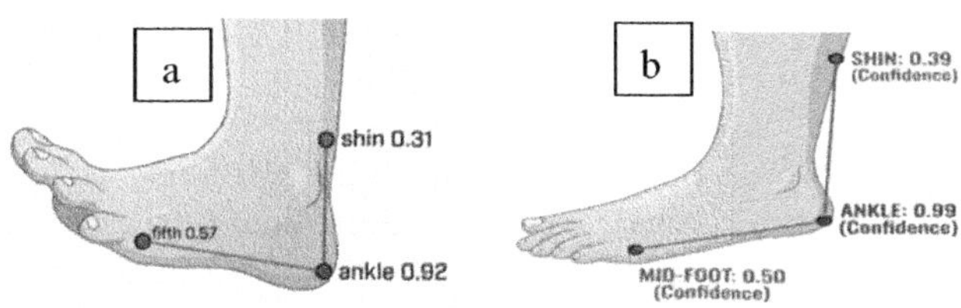

Fig. 1. Representative system output. Colored markers indicate the three automatically detected anatomical key points (tibial midline, lateral malleolus, fifth metatarsal head); Confidence scores are shown adjacent to each key point. (a) Dorsiflexion (b) natural

Table 2 reports the full evaluation results for the best snapshot (iteration 120).

Table 2. DeepLabCut model evaluation

Metric	Train	Test
RMSE (pixels, all predictions)	6.07	8.05
RMSE (pixels, at p-cutoff)	5.27	7.52
mAP (%)	97.54	94.48
mAR (%)	98.41	96.76

The results demonstrate strong model performance. The test mAP of 94.48% and mAR of 96.76% confirm that the model generalizes well to unseen images. The p-cutoff RMSE of 7.52 px on test data reflects localization accuracy for predictions that the model is confident about, and compares favorably with the full-set RMSE of 8.05 px, indicating that the confidence filtering mechanism effectively separates reliable from uncertain predictions. The train/test gap in full-RMSE (6.07 vs. 8.05 px, a ratio of 1.33×) is modest and does not indicate substantial overfitting, which is consistent with the dataset size and the regularization provided by DeepLabCut's data augmentation pipeline. The p-cutoff RMSE values (5.27 train, 7.52 test) are the operationally relevant metrics: they represent the localization error for predictions that the system accepts and acts on, and both fall within the sub-10-pixel range that yields clinically acceptable angle estimates.

To contextualize the clinical relevance of the p-cutoff RMSE, we derive the implied angle error from the imaging geometry. The iPhone 14 rear camera has a horizontal field of view of approximately 77°, which across the 4032-pixel horizontal dimension yields an angular resolution of $77° / 4032 \approx 0.019°$ per pixel. At a camera-to-subject distance of 1.2 m, one pixel corresponds to a physical displacement of $1.2 \times \tan(0.019°) \approx 0.40$ mm. A test p-cutoff RMSE of 7.52 px therefore corresponds to a landmark localization error of approximately 3.0 mm.

Propagating this localization error to the angle computation: for the foot vector (lateral malleolus to fifth metatarsal head, anatomical length approximately 100 mm in the study population), the implied angle error is $\arctan(3.0 / 100) \approx 1.7°$. For the shank vector (approximately 150 mm), the implied error is $\arctan(3.0 / 150) \approx 1.1°$. Both values fall well within the $\leq 5°$ clinical requirement, even under the conservative assumption that localization error affects both endpoints of a vector simultaneously. These anatomical lengths are population averages for the study cohort; the planned clinical validation study will provide direct empirical confirmation of the angle error against goniometric reference measurements.

4.6 Six-Stage Processing Pipeline

The complete system pipeline consists of six stages: (1) dual-image capture (neutral and maximum dorsiflexion); (2) image upload via the web interface and REST API to the backend server; (3) DeepLabCut inference using the ResNet-50 model, returning three key point coordinates with per-key point confidence scores; (4) confidence-filtered key point selection with uncertainty flagging for sub-cutoff predictions; (5) vector-based angle computation and ROM delta calculation; and (6) standardized digital report generation with key point overlay visualization, the computed angle, and optional session logging for longitudinal tracking.

The modular architecture (Python/Flask backend, DeepLabCut inference module, React frontend) allows independent improvement of each component. The system is designed to process a complete measurement session in under a minute from image submission to result display.

4.7 HCI Design for Clinician-in-the-Loop Deployment

The interface design is grounded in four principles derived from the assessment of needs. Minimal interaction cost: no specialist equipment, no markers, measurement cycle under a minute [11]. Transparent landmark visualization: key point overlays are displayed before the angle is accepted, allowing the clinician to verify or reject the model's localization, preserving clinical judgment rather than bypassing it [13]. Longitudinal tracking: optional session logging supports rehabilitation progress monitoring across visits [12]. EHR integration: the report output is formatted for direct entry into documentation workflows [2].

5 Planned Clinical Validation Study

The planned study evaluates the system against two reference measurements simultaneously: the universal goniometer (clinical gold standard) and a custom protractor-mounted leg-placement device that provides an independent mechanical reference. The leg-placement device consists of two wooden pieces connected with a hinge, ensuring a 90-degree baseline ankle angle; the leg is secured with Velcro straps while the foot dorsiflexes freely. This dual-reference design allows us to separately quantify error against the goniometer and against a mechanical ground truth, providing a clearer picture of the system's accuracy.

Participants will be healthy adults aged 30–70 years with no musculoskeletal disorders or ankle injuries at the time of testing. Each participant will perform several dorsiflexion repetitions per session; all three measurement methods are applied simultaneously to each repetition. Primary metric: mean absolute error (MAE) of the AI application relative to the goniometric reference. Secondary metrics: p-cutoff key point detection success rate, processing time per session, and first-time user completion time.

6 Discussion and Rehabilitation Implications

The model evaluation results establish that the DeepLabCut ResNet-50 system reliably detects ankle key points with test mAP of 94.48% and p-cutoff RMSE of 7.52 px. The confidence-filtering mechanism is a clinically important design choice: by making uncertainty explicit rather than reporting all predictions unconditionally, the system supports trustworthy deployment in settings where an incorrect angle could lead to a harmful clinical decision. This aligns with findings in the literature that clinicians distrust AI tools that produce results without explanation [13].

For aging populations, the implications are particularly significant. Mobility decline in older adults is gradual, often sub-threshold for clinical detection until it has progressed substantially. A tool that takes two photographs and reports a ROM measurement, with longitudinal logging, fundamentally changes the feasibility of regular ankle ROM monitoring outside of formal clinical assessments. The lightweight acquisition requirements also make the system viable for remote monitoring scenarios, where patients capture images at home, transmitting results to a clinician for review [4, 7].

The train/test RMSE gap reflects the limited size of the current dataset, which is a common constraint in clinical AI development [16]. The ongoing data collection in the clinical validation study will expand the training set and is expected to reduce this gap. The modular architecture allows the model to be retrained and improved without changes to the angle computation or interface layers.

Limitations include small training dataset; validation currently limited to healthy young adults; no multi-planar measurement (plantarflexion, inversion, eversion are identified as future work by survey participants); and the planned validation study has not yet been completed. These limitations bound the current claims and define the next development phase.

7 Conclusion

This paper presented an AI-based markerless goniometer for ankle dorsiflexion measurement, combining a clinician needs assessment with a trained and evaluated deep-learning system. The needs assessment documented a widespread abandonment of formal ROM measurement in clinical practice, driven by the goniometer's poor reliability and procedural burden, and identified concrete downstream clinical consequences for patients in rehabilitation and aging care. The AI pipeline, based on DeepLabCut with a ResNet-50 backbone, was trained on a dedicated clinical ankle dataset and achieved a test mAP of 94.48% and test p-cutoff RMSE of 7.52 px at the best snapshot (iteration 120). The confidence-filtering design makes uncertainty explicit, supporting safe clinical deployment. A clinical validation study comparing the system against the universal goniometer and a mechanical reference device is underway. The system represents a concrete step toward accessible, reliable, and longitudinal ROM monitoring in rehabilitation and aging care.

Declaration on Generative AI. During the preparation of this work, the authors used GPT-4 and Claude (Anthropic) for grammar and spelling checking, manuscript structure, and literature synthesis. The authors reviewed and edited all content and take full responsibility for the publication. All research design, data collection, analysis, and interpretation are the authors' original contributions.

References

1. Arnold, B.L., Wright, C.J., Ross, S.E.: Functional ankle instability and health-related quality of life. J. Athl. Train. **46**(6), 634–641 (2011)
2. Asan, O., Alparslan, E., Choudhury, A.: Artificial intelligence and human trust in healthcare: focus on clinicians. J. Med. Internet Res. **22**(6), e15154 (2020)
3. Brockett, C.L., Chapman, G.J.: Biomechanics of the ankle. Orthop. Trauma **30**(3), 232–238 (2016)
4. Bok, S.K., Lee, T.H., Lee, S.S.: The effects of changes of ankle strength and range of motion according to aging on balance. Ann. Rehabil. Med. **37**(1), 10–16 (2013)
5. Cai, C.J., et al.: Human-centered tools for coping with imperfect algorithms during medical decision-making. In: Proceedings of the CHI 2019, pp. 1–14 (2019)
6. Collins, N., Teys, P., Vicenzino, B.: Reliability of ankle dorsiflexion measurement using different clinical techniques: a systematic review. J. Foot Ankle Res. **7**, 21 (2014)
7. Goes, S.M., et al.: Ankle dorsiflexion may play an important role in falls in women with fibromyalgia. Clin. Biomech. **30**(6), 593–598 (2015)
8. He, K., Zhang, X., Ren, S., Sun, J.: Deep residual learning for image recognition. In: Proceedings of the CVPR 2016, pp. 770–778 (2016)
9. Kidzinski, L., et al.: Deep neural networks enable quantitative movement analysis using single-camera videos. Nat. Commun. **11**, 4054 (2020)
10. Litjens, G., et al.: A survey on deep learning in medical image analysis. Med. Image Anal. **42**, 60–88 (2017)
11. Mathis, A., et al.: DeepLabCut: markerless pose estimation of user-defined body parts with deep learning. Nat. Neurosci. **21**(9), 1281–1289 (2018)
12. Ota, S., Ueda, M., Aimoto, K., Suzuki, Y., Sigward, S.M.: Acute influence of restricted ankle dorsiflexion angle on knee joint mechanics during gait. Knee **21**(3), 669–675 (2014)

13. Roggio, F., et al.: A comprehensive analysis of the machine learning pose estimation models used in human movement and posture analyses. Heliyon **10**(21), e39977 (2024)
14. Rome, K., Cowieson, F.: A reliability study of the universal goniometer, fluid goniometer, and electrogoniometer for the measurement of ankle dorsiflexion. Foot Ankle Int. **17**(1), 28–32 (1996)
15. Unkelos-Shpigel N., Kroitoro R., Zisman S., and Ravid E.: Toward an AI-based ankle goniometer: a clinician needs assessment. In: INI-DH 2026: 2nd Workshop on Innovative Interfaces in Digital Healthcare, in conjunction with the 2026 International Conference on Advanced Visual Interfaces (AVI 2026), 08–12 June 2026, Venice, Italy (2026)
16. Wekenborg, M.K., Gilbert, S., Kather, J.N.: Examining human-AI interaction in real-world healthcare beyond the laboratory. NPJ Digit. Med. **8**(1), 169 (2025)
17. Yang, Q., Steinfeld, A., Rosé, C., Zimmerman, J.: Re-examining whether, why, and how human-AI interaction is uniquely difficult to design. In: Proceedings of the CHI 2020, pp. 1–14 (2020)
18. Youdas, J.W., Bogard, C.L., Suman, V.J.: Reliability of goniometric measurements and visual estimates of ankle joint active range of motion. Arch. Phys. Med. Rehabil. **74**(10), 1113–1118 (1993)

Author Index

S. S. Khan et al. (Eds.): IJCAI 2026, CCIS 3127, p. 175, 2027.
https://doi.org/10.1007/978-981-92-5190-2